William Glenny Crory

Plain directions for practical farmers on flax culture

William Glenny Crory

Plain directions for practical farmers on flax culture

ISBN/EAN: 9783742845115

Manufactured in Europe, USA, Canada, Australia, Japa

Cover: Foto ©berggeist007 / pixelio.de

Manufactured and distributed by brebook publishing software
(www.brebook.com)

William Glenny Crory

Plain directions for practical farmers on flax culture

PLAIN DIRECTIONS

FOR

𝔓ractical 𝔉armers

ON

FLAX CULTURE.

BY

W. GLENNY CRORY,

AUTHOR OF INDUSTRIAL RESOURCES (STILL NEGLECTED) IN IRELAND
INDUSTRY IN IRELAND, ETC ETC.

Second Edition.

DUBLIN: WILLIAM ROBERTSON, 23, UP. SACKVILLE-STREET.
LONDON: SIMPKIN, MARSHAL AND CO.
EDINBURGH: JOHN MENZIES

1864.

CONTENTS.

Printed by R. CHAPMAN, Dublin.

PLAIN DIRECTIONS

FOR

FLAX CULTURE.

MANY farmers who have not sown flax, suppose it to be a crop either very difficult to manage, or dangerous in a pecuniary sense, or so likely to be a failure, that it ought to be avoided altogether. Under these circumstances, before stating a few rules for the management of the crop, I shall endeavour to disabuse the minds of Irish agriculturists of certain false ideas they have entertained, and do still cling to, in regard to flax culture. I have no hope, however, in succeeding with any one now opposed to Flax culture who is not disposed to respect evidence, or who is not ready to be guided by common sense. The grand principle I advocate is profit, and the crop that I commend is the one that will pay best. But before any one can benefit by what I am about to say, the mind must be freed from the following prejudices, viz. :—

First.—That flax is more exhausting on the soil than any other crop. The experience of those who have grown it, the testimony of those who have examined soils before and after flax crops had been raised on them, and common sense itself, all combine to say that in no peculiar sense, if in any sense at all, is flax an exhausting crop.

FLAX will not exhaust ordinary soils more than wheat, oats, or barley; at the same time, if flax be sown on very poor soil, those elements on which the fibre feeds being extracted, may leave the land in a state seemingly more prostrate than is usual after most other crops commonly cultivated in this country. I have often found that persons who chanced to grow flax,

A 2

but who also allowed their lands to become exhausted through neglect, have laid the blame on that crop rather than take it to themselves. This accounts for the prejudice in many cases.

Secondly.—It is a great mistake to suppose that the cultivation of FLAX necessarily leads to a scarcity of manure. The fact is, if flax be properly managed, the portion of a farm devoted to that crop returns as much manure, although not quite so directly, as can be got from any other part. If the bolls be carefully rippled, and used for feeding cattle or calves, the result with regard to manure will be as valuable as using either wheaten or barley straw for litter, or oaten straw for fodder, can possibly be. Besides, if the shoves from the scutching mill be put into the pool in which the flax had been steeped, a quantity of very superior manure will be obtained, sufficient to repay all the extra labor requisite for this economy. But if this be inconvenient, there are other ways of raising manure from flax crops in as great quantities as can be got from any crop that could be sown instead of flax. For the sake of manure it would be well worth the cost and labour to bring home both long and short shoves for putting into pigs' beds, or for being used as an absorbent of excessive moisture in a farm-yard and the area of a piggery. Nor are these the only ways by which flax crops may become the occasion of increasing the supply of manure. If in cases where the flax is not scutched but kept over winter, the water of the steep-pool were absorbed by peat mould, or any one of the many other dry substances so plentiful about a farm-yard, the water so impregnated with the flax oil, would, so soon as it would be applied to the field in which the flax was grown, become not only as compensative as anything to be had from other crops, but very much more so. Perhaps the origin of this sadly too prevalent notion of flax crops being against manure gathering, may be traced to ignorant or idle persons who wished to find excuses for not growing flax, and as manure is to the farmer the most important element in good farming, next to capital and industry, any thing supposed to lessen the supply of that ingredient would be most likely to alarm all who would not take the trouble to examine the case for themselves. Were it justifiable in a work of this sort, or necessary for any good reason, I could show by calculations, which no practical farmer could disprove, that flax crops rather tend to increase than to diminish manure. But in this case, as in many others, before I could deal fairly with the matter, I must be allowed to assume that farmers are prepared to amend their ways by throwing off their indolent habits in respect to manure gathering in general. I do admit, however, there are some farmers who will never either economise the shoves, or use the steep-water of their flax; but their conduct can in no sense be considered a specimen of that of Irish farmers generally. It is unfair, moreover, for farmers to make out against flax crops a deficiency of manure, when the seeming difference in favour of using the straw of an acre of grain for bedding is attributable to the fact that it has only to be brought to the stable or byre, put under the beasts to insure its finding its way to the dunghill, whereas the manure-producing parts of a flax crop must be looked after as carefully as the fibre or the seed, and is in fact, in its own place, and to the extent of its money value, as important as either.

Thirdly.—Another objection, originating in ignorance of the crop, is, that its adoption reduces the supply of food on a farm. At first sight it seems as if this were a plausible objection. But no sooner is it put to the test of common sense than it departs, for want of a single fact in all the experience of flax growers to support it. If this complaint against flax culture were well founded, it implies that the admission of flax crops into rotation necessarily reduces the quantity of grain raised on a farm. It is well known, however, that no more that one-fourth, one-fifth, or one-seventh portion of a farm is ever likely to be devoted to flax. Experience also goes to prove that those farmers who are careful to give a fair space to flax, are always best able by increased profit in that crop to pay for manure and labor, for the portion under corn; and the result is, that in this mixed system of culture there is often increased supplies of food.

Taking the whole area under crops in Ireland at 6,000,000 of acres in round numbers, the flax crop of 1863, although 60,000 acres in excess of that of 1862, would be very little more than one-thirtieth of the whole cultivated area; and yet this one-thirtieth part would displace no more than £2,250,000 worth of food at the highest calculation, although it would yield for fibre alone £3,500,000, at the lowest estimate, after deducting from the value of the flax crop the extra cost of labor, as compared with grain crops. This £1,250,000 of extra income would provide a supply of food for men, and fodder for beasts, which would enable farmers, of any grade, to live quite as well themselves as gentlemen farmers now live, and to keep their live stock in as good condition as the like are kept on the largest farms. Besides, and more particularly the sowing of a fair proportion of the cultivated land, with flax every year, would become the means of putting money into farmers hands, sufficient to allow them to eat more of the corn crops, pigs, fowl, and other produce raised on the farm.

I have not been able, after a careful review of the various arguments for and against flax culture, to come to any other conclusion, than, that on the whole, flax-growing is sufficiently simple in all its branches, to be carried out by people in any part of the country, and if cultivated in a fair relation to other crops, growing it is the best way to make the most of a farm, whether it be of 10, 20, 50, 500, or 1,000 acres, irrespective to a large extent, moreover, of whether the farm be in Connaught, Munster, Leinster, or Ulster;—the best flax growing lands, being in Connaught and Munster, and the worst in Ulster and Leinster. The following extract is from " A REVIEW OF THE IRISH LINEN TRADE OF 1863," and has my entire approval, except that I would be much inclined to fix the number of acres of flax which ought to be cultivated at twice, if not thrice the numbers named.

"FLAX—*The area under flax in 1863 was 214,092 acres, exceeding the largest acreage of any previous year by 39,513 acres; the previously largest crop was that of 1853—say 174,570 acres. In 1862 the acreage under flax was 150,070 acres, thus showing an increase of say, 43 per cent., comparing the acreage of 1862 with that of 1863. The yield this season, though not realising the early anticipations of many, shows great improvement as compared with that of the preceding year. The very best-informed parties in this district presume on the produce being, say, 60,000 tons, the one-half of which quantity may be said to have been already disposed of, either in open markets or at the various scutch-mills throughout the country.* IT WOULD BE QUITE IMPOSSIBLE TO USE ANY LANGUAGE THAT WOULD BE EXTRAVAGANT IN SETTING FORTH THE ADVANTAGE, IN A MERCANTILE AND NATIONAL POINT OF VIEW, OF THE EXTENDED CULTIVATION OF THE FLAX CROP IN IRELAND. *It is only necessary to state the fact, that a sum not less than £4,500,000 shall have been expended in the purchase of flax throughout the country, when the season terminates. The advantages likely to result from the distribution among the farmers and agriculturists of such an enormously large amount of money are so obviously apparent that it would be little short of an impertinency to point them out, or to chronicle them in detail. To a sense of these advantages the people of this country seem at length to have become awakened. The hope now is not an idle one, that, in the ensuing season, an area in Ireland of 300,000 acres will happily be found devoted to flax production."*

Since writing the first Edition, I have had the pleasure of receiving several valuable criticisms from many flax growers; in reference to all which I have to say that, having considered them, I admit that the quantity of seed I have prescribed is too much for some very rich well tilled lands, such as are in some parts of Ulster, but scarcely enough for some of the very poor lands in Connaught; for the average of soils, however, I still believe the quantities I have given are correct. I am indebted to a Landed Proprietor in the County Down, who has fully satisfied me by facts, that in wet weather the less the ground is worked, either before or after the seed is sown, the better.

CULTIVATION OF FLAX.

SOIL.

In the selection of soil for sowing flax the very best rule would be to observe no rule at all. However paradoxical this may seem, it is nevertheless the best way of disposing of a very critical question. In a climate like that of Ireland there is not a square mile of arable land in which flax may not be grown successfully, provided the land is drained, the field is ploughed, cross-ploughed, harrowed, cross-harrowed, and properly prepared by breaking the clods for the seed. In most directions I have met with on flax culture, it seems to be assumed that each farmer has such a variety of soils at his disposal that he can choose whatever kind he pleases for his flax crop. Also, that this may be done without any interference with his oat, wheat, turnip, or other crops. This error is easily accounted for by the fact, that most of those to whom agriculturists are indebted for such works, are either chemists too much engaged in scientific labors to have visited farmers at their homes, or writers whose object is rather to produce a readable book than to promote practical farming; most directions, therefore, already extant, bear on their face the results of a straitened process of induction, a limited circle of inquiry, and also too great a desire to please, and too little to reform those for whose instruction they are written. I mention these facts, partly by way of criticism, and partly in order to preface the announcement of a different, and I flatter myself—a better mode of treating the subject. What the practical farmer wants to know, is not so much the best soil for growing flax, as whether or not he is likely to have in his own farm, be it large or small, any ground suitable for flax. In dealing with the subject in this way, I come to the root of the matter; and it gives me much pleasure to say, that without exceptions to any extent worthy of notice, the following is the order in which the best soils for flax-growing may be arranged, including all parts of the United Kingdom. IRELAND has not only a greater proportion of flax soils than England, but climate considered, which is for all practical purposes, inseparable from soil, she has more FLAX GROUND than is to be found in all England, Wales, and Scotland.

In Ireland all kinds of fibre from the coarsest to the finest, may, in the average of any seven seasons, be profitably produced; whereas, in such parts of England as produce the best peas, in hop districts, and in cheese-raising soils of other parts, as well as in the withering atmosphere of iron districts, flax, except rarely, is not likely to be a paying crop. I believe, moreover, that there is more than accidental connection between the two facts, that where butter is most easily made, there flax is most profitably grown. If, however, the English farmers, in other parts outside those indicated, devoted attention to the rearing of seed, their earlier seasons and better system of labor would unquestionably give them advantages over Irish farmers, but certainly not over Scotch farmers, in this line ; for it is a fact worthy of observation, that the flax seeds found on the whole to be the best are raised in higher latitudes than those in which they are sown. I merely mention these facts in anticipation of the probability of alarm seizing the farmers of Ireland when they hear of their Scotch and English neighbours having also (as I am sure they soon will) taken to flax culture. But as respects the climate, and to some extent the soil as well, Ireland has advantages over England and Scotland, as great in regard to flax crops as these have over Ireland, in respect to wheat, peas, barley, cheese, hops, and farm-seeds. The improved and improving system of agriculture in Great Britain, must not be overlooked in its striking contrast with the absence in Ireland, even allowing for recent reforms, of the merest necessities of decent husbandry in very many parts of the west, some parts of the south, a few places in the midland districts, and even here and there in the north. Nevertheless, Flax has been grown successfully in this country—in bogs, on the sides of mountains, in open, level plains—in fact, near the roar of the Atlantic, as well as on the shores of the Irish sea ; and by all classes of farmers except sluggards, the holders of undrained swamps, or badly-tilled fields, allowance being made for even bad implements. The best soil for flax is a loam. The best condition is dryness, and if deep so much the better, and best of all a clay subsoil. An equally particular description might be fixed on soil for any other crop. But so well convinced am I by the strictest inquiry extending over every part of the island, that success depends, not so much on the character of the soil as its condition, and not so much the place flax has in a rotation, as the attention it gets from the cultivator, that I repeat the remark I have already made, that for flax-growing there is no special necessity for a particular soil.

I wish, however, to be understood exactly, for although I have not found in any quarter a particle of evidence to justify any difficulty in the selection of a field to sow flax in, I have also discovered wherever I go, that great importance attaches to the selection of soils for every crop—flax included. It is just as great a mistake to sow wheat, oats, or barley, or to plant potatoes in an unsuitable field, as it is to misplace flax ; for notwithstanding the popular notion, that any farmer may know the right field for every other crop by intuition, and that it is requisite he should have special instruction in regard to flax, I say, that although I did not expect to find it, yet I have good reason to state that Irish farmers are farther behind the age in the culture of cereals than in the management of flax. Many farmers in Ulster have succeeded in flax, whilst those in the other provinces have not of late even fairly tried it. Above all the other causes, however, which brought about this anomaly, the separation of the flax from other crops, and the attributing to its culture a series of difficulties, which only exist in the imagination, have been most effective in raising up hindrances to the growing of flax, although it is the best paying crop Irish farmers have as yet cultivated. I fully approve of SPECIAL INSTRUCTORS for the management of flax, especially in Connaught, Munster, and Leinster; but I do not admit that there is one iota less necessity for a like course being adopted in regard to every other crop—potatoes included. By correctly selecting soils for other crops, the right field is sure to be allotted to flax, although judicious selecting for flax merely may not ensure other crops a prosperous place.

I have learned, moreover, that even in Ulster random and sometimes rotation, have become guides much more than the rules laid down by chemists on the subject of soils. I believe it is better that it has been so, for until the practical farmer generally becomes an analytic chemist, it is dangerous for him to go by the crude notions a smattering of that science may lead him to adopt. Small farms do not admit of that strict adherence to scientific farming which may be carried out on large farms. But this is neither the only nor the greatest difficulty in applying rules for the selection of soils founded on the analysis of crops. Some years ago Sir Robert Kane made, at the request of the Flax Improvement Society of Ireland, analyses of "soils which had produced remarkably good crops of flax," which soils were found to be composed of, viz.:—(vide Industrial Resources, p. 325.)

	No. 1.	No. 2.	No. 3.
Silica and Silicious Sand	73·72	69·41	64·93
Oxide of Iron . .	5·51	5·29	5·64
Alumina	6·65	5·70	8·97
Phosphate of Iron .	·06	·25	·31
Carbonate of Lime .	1·09	·53	1·67
Magnesia and Alkalies, with traces of Sulphuric and Muriatic Acids .	·32	·25	·45
Organic matters .	4·86	6·67	9·41
Water	7·57	11·48	8·73
	99·78	958	100·11

Now, in relation to these analyses, three questions must be answered before recommending them as guides for practical farmers. First: Are these soils the best in all Ireland for flax growing? If it be answered. They are, which is probably a correct reply; then comes a second question, namely, in what proportion do these soils stand to all the soils of the country; or, rather, how many acres of the average of these three kinds could be had all over Ireland? Before answering this question it must be asked, thirdly, how is the acreage to be ascertained, and also, in cases where there are several soils in the same farm, and sometimes more than one in the same field, how is that which each farmer has, to be brought to this standard, which from the character of Sir R. Kane, I freely admit is as likely as others which differ from it, to be the best that could be adopted.

If farmers must wait till the soils of their farms can be analysed by eminent chemists, they will have to wait a long time, even if each had a sufficient fee ready, and could find a competent practitioner waiting to accept it. Still less likely is it that a mere chemical test will ever be supplied, if amateurs are allowed to be judges. Therefore, practice demands that in regard to flax culture, just as in respect to all other crops, farmers should avail themselves of those rules which experience suggest, except when they feel justified in making new experiments, or see their way clearly to reforms, enforced by sound theory, which always requires to be tested.

Chemistry is as much indebted to observation for checking and regulating the results of analyses, as rules made by farmers are to chemistry for correcting errors which attach to their application. But the error to which the Irish farmer is unhappily most of all prone, is the disregard of both chemistry and custom ; or, what is worse, such a mixing up of the two, without judging of either on merit, as defeats the object of both. From a careful examination of all the facts I can gather relating to the selection of soils for flax, I arrive at the following inferences :—

First.—That there is a fair chance of a good flax crop on the average kinds of soil to be found in every part of Ireland ; also that whilst the climate of Connaught and Munster, with parts of Leinster, is more characterized by the " half sunshine half shower" peculiarity of Ireland (which is scarcely humidity) than is found in most parts of Ulster, the fact of Ulster being the flax growing province is a gross anomaly.

*Secondly.—That the only way to enable the Irish farmer to select soils correctly, is for him either to learn chemistry, in so far as it relates to soils, thoroughly, or to cause his son or sons to do so. By "*LEARNING*" I do not mean merely attending lectures, however able, but I mean becoming the pupil of some able chemist, and being taught not only to analyse, but instructed in the principles of the science. Until this* *is the case, all Irish farming will be alike badly done, but other crops will suffer just as much as flax.*

Thirdly.—I also infer that the proportion of soil in Ireland suited to the several ordinary crops is so large, that by a seven years' rotation no less than ONE MILLION *of acres out of lands now reclaimed would be devoted to flax ; and also that as waste lands are reclaimed, and the area of arable ground enlarged, the number of acres suitable for flax would be more than proportionately increased ; because many parts of several farms which, for want of drainage, cannot now be used for flax crops, under such reforms as the reclamation of waste land implies, would be made into good flax soil.*

Fourthly.—I infer from all I can learn that many persons are deterred from flax culture by the advice of landed proprietors, gentry, and others on whose judgment they rely ; and I must be excused recording the disagreeable fact that many who take upon them to give advice, are utterly incompetent to do so—they are either prejudiced against what they have never taken the trouble to investigate or are misled by persons on whom they rely ; or what is a more common evil (and peculiar to Ireland), is, that noblemen, landed proprietors, and speakers at agricultural meetings often receive and frequently repeat what persons (so called practical) tell them, whilst at the same time these persons know they are expected to say only what will please and take care to do so. In fact this evil is carried often to so great an excess that well-informed but temporising men are intimidated or enticed into giving opinions exactly contrary not only to truth but to their own convictions. Evils arising from these and kindred causes have kept millions of money out of the pockets of Irish farmers who, suffering from these consequences, sometimes administered direct, but often through societies, have been kept from cultivating flax crops.

PREPARATION OF THE SOIL.

PREPARATION of the soil for the seed comes next after the selection of soils; but whilst I have felt it best to lay down no fixed rule for that part of the culture, I feel that too great attention cannot be given to specific rules on this point. The question of ROTATION ought, perhaps, to come first. I really cannot learn, however, that there is either any general agreement on this matter, or that it has ceased to be an undecided point, if not a *puzzle* to farmers.

As regards flax, however, which is all that concerns me, I believe it may be grown now as it often has been immediately after potatoes, and for several years in succession on the same field. The seven-year rotation is, on the whole, the best for Flax crops. Next after an Oat crop sown on lea, or next after a white crop sown after a root crop, is the best time for Flax. Sowing on lea without the intervention of another crop is questionable, if not dangerous ; and although it might be done, provided the ground is very carefully prepared, it is, nevertheless, unadvisable, because Flax ground to a sufficient extent may be had without any such extraordinary proceeding.

Ground in good heart does not require special manuring for Flax. But if Manure is used at all it ought to be specially prepared for the flax crop, unless it be applied for the purpose of raising the general fertility of the soil, in which case, Lime may be applied, but not less than six weeks before the seed is sown. Guano may also be used, but Flax manure if skilfully compounded, and judiciously applied, may be of greater use in most cases than any guano. The several manufacturers of Artificial manures have produced good Flax manure, which has in places where it was required, done much to improve the yield, and also to give vigor to the young stalks, which told in the end in favor of the quality of fibre. It would be unfair to mention any one by name, except to correct a popular notion in the South and West, that none but Northern manufacturers understand how to make Flax manures, and also to gain for manures made up by prescription their proper place beside Guanoes, which, valuable as they are, may be and have been equalled by the Dublin manufacturers, as well as by the eminent firms in Belfast, who are so justly celebrated for the manufacture of Flax

Manure. The Iodine and Marine Salts Company, of Galway, propose to make Manures; and from the peculiar nature of the debris of Sea Weeds, which will form the basis of their compost, they are likely to make excellent Flax manures.

Land intended for Flax ought to be ploughed in autumn and cast up in flats, to catch the effects of winter. In spring there ought to be another ploughing four to six weeks before sowing the seed, so as to allow weeds to come up that they may be rooted out by the harrowing of the ground before the sowing. If the ground is ploughed sufficiently, the roots of the flax will penetrate directly to a very considerable depth. The great depths to which the roots go in light friable loams, accounts more than any other consideration for the better crops of Flax on those kinds of soil than on any other. The more stiff the ground the greater necessity there is to bring it to a condition in which the early tender buddings may get through the earth without being bent, and the roots get down without being stopped. Ploughing, breaking, harrowing, pulverising, carrying off stones, and every other necessary work must be done until the ground is clean and level. Every farmer would require a special set of rules—each farm ought to get the peculiar treatment suited to its condition; and, in fact, in the same field there is often several kinds of soil, each requiring to be dealt with to some degree differently. In a general way, however, any farmer will know how to bring his land to the condition of an ONION BED, which is very fairly considered a proper pattern for a Flax field. A light rolling is often useful to make an even surface before the seed is sown. This ought to be done after the last harrowing, and then broken up by the seed-harrow immediately preceding the sowing. The depth and number of ploughings, and, in fact, most of the operations must be modified by the nature of the soil, and also, and more particularly, by the absence or existence of thorough drainage. When land is not drained (which is the greatest of all drawbacks on Flax culture) the autumn ploughing ought to be at least five to seven inches, and the spring ploughing four to six inches, deep.

THE SURFACE SHOULD BE MADE EVEN, AND THE RIDGES WITH A SLIGHT ELEVATION IN THE CENTRE, WHICH OUGHT TO BE MORE OR LESS ACCORDING AS THE SEASON PROMISES TO BE DRY OR THREATENS TO BE WET. THE PLOUGHING AND HARROWING OUGHT TO LEAVE THE GROUND STRAIGHT, LYING IN LINES AND NOT ZIGZAGGED. ALL CLODS OUGHT TO BE CAREFULLY BROKEN AND NOT ONE LUMP LEFT TO BAFFLE THE EARLIER EFFORTS OF THE YOUNG PLANT AT COMING UP TO COVER THE GROUND.

If these regulations be attended to, WEEDING will be unnecessary, but in cases were weeding cannot be avoided, let it BE ALWAYS DONE BY PERSONS OF LIGHT WEIGHT, MOVING IN THE SAME DIRECTION, AND FACING THE WIND.

It ought not to be overlooked, however, that weeding is more or less necessary, in proportion as the ground is properly prepared, and may be avoided altogether by a little timely care.

SEED.

SEED comes next, and above all other desiderata GOOD SEED and skilful sowing are indispensable to profitable flax crops. Riga Seed is better than Dutch, and either is preferable to American. Home grown seed has neither been produced to a sufficient extent, nor as yet fairly tried, to enable one to find out data for an opinion worth recording as to its merits. Excellent crops have frequently been raised when it has been sown, but instances of failure are also on record, on which the whole blame is laid on the home-saved seed. English seed is also hardly spoken of by many farmers. Failure might, however, have followed Riga seed in those cases when it happened with Irish or English seed.

There is no doubt that Irish Seed growers do not take pains in saving the seed carefully. But there is no reason why they might not do so, and if they did, I hesitate not to predict that Irish grown Seed would be as good as Riga. The only thing I see in favor of Riga seed must be accounted for by the management the growers apply to it, for the difference in latitude could not make any difference so great as that which exists in favor of Riga over Irish Flax seed.

The rearing of Seed for sowing is one of the branches of Flax culture, which ought to be attended not only for the profits it would yield, but also as a means of checking the practice of selling old or mixed seed for new and good.

In 'selecting seed, the greatest attention should be given to the following rules. IN ALL CASES EVERY FARMER OUGHT TO BE SATISFIED THAT THE SEED HE SELECTS IS NEW, AND HE OUGHT ALWAYS TO BUY THE BEST, REGARDLESS OF PRICE. FOR ALL KINDS OF LANDS, RIGA SEED WILL DO ; BUT IF DUTCH BE SOWN, IT OUGHT TO BE IN HEAVY OR BOGGY SOILS. AMERICAN SHOULD NEVER BE SOWN IN ANY BUT VERY LIGHT, SHALLOW GROUND.

New seed is always shining to the eye, slippery in the hand, and feels plump and heavy. There is but one way, however, to be assured of good seed, so long as foreign is used, and that is to buy from none but merchants of the highest respectability, who are above the crime of deception, be the pecuniary temptation what it may.

SOWING.

Before entering on this part, I wish to call attention to the following quotation from a report of a speech made at a very interesting and meeting lately held under the patronage of the Earl of Erne, whose object in convening the meeting was, to bring the subject of flax culture before his tenantry. The gentleman who made the following remarks is Mr. Weir, Lord Erne's agriculturist. The statements are not only on the whole valuable in themselves, but having been widely circulated and greatly respected, I wish to give my readers the benefit of my opinion of them. Besides, what Mr. Weir has said is a fair specimen of what agriculturists of his class generally believe although I consider some parts of his observations open to serious objections.

"The best soil to grow flax on is sound, dry, deep loam, on a clay subsoil. If not naturally quite dry, it is necessary that it should be perfectly drained, and if subsoiled, it should have been done at least two years before sowing flax on it, as flax requires an evenly and rather firm subsoil for the roots to grow in. Gravelly lands are not suitable for flax; neither are light, boggy holms: in wet seasons fine flax is often produced on dry, sandy lands. The best state of the land for flax, is either after lea from which one crop of oats has been taken, or after a grain crop which has followed a crop of well cultivated and well manured potatoes or turnips. The rotations then may be,—Where flax follows a grain crop after lea, first year, potatoes and turnips; second year, grain crop with grass seed and clover; third year, hay and soiling; fourth year, pasture; fifth year, oats; sixth year, flax. Where flax follows a grain crop after potatoes and turnips—first year, potatoes and turnips; second year, grain crop, wheat or oats; third year, flax with grass seed and clover; fourth year, hay with soiling; fifth year, pasture; sixth year, oats. The seventh year commences the same rotation again. If thought desirable, the land in either of the above cases might get another year in grass; and if so, seven fields would be required to complete the rotation in seven years. If the land has been previously properly cultivated, two ploughings will be sufficient; but if very weedy, three may be necessary to get it into proper state for the seed. The first ploughing should be given in autumn, immediately on the removal of the grain crop. If the land is damp, it should be ploughed in ridges and water-furrowed, to keep it from being too much saturated in winter. In the month of March following, it should be well harrowed, and again ploughed and harrowed, and if scutch or other root weeds come up, they should be picked off. If the land is in proper condition, no direct application of manure is necessary for this crop; but it is considered useful to lime the land for the previous green crop, or to top dress the lea with lime compost some time previous to breaking it up for the grain crop, and three or four cwts. of kelp ashes per acre has been sometimes applied to the flax crop with beneficial results. Flax may be sown any time during the month of April, or even the first week of May; but the second and third week of April is the proper time. Riga seed is the best kind to sow on most soils, and the quantity usually given is 33 old Irish gallons to the Cunningham acre; this is about 40 1-7th old Irish gallons to the Irish acre, or 24¾ of the same gallon to the English acre. Immediately previous to sowing, the land should be harrowed until perfectly pulverized; after sowing, a double turn of light grass seed harrow is sufficient to cover the seed. It should then be finished with a light roller; if the land and weather is very dry, a heavier roller may be used. If proper attention has been paid to the previous cultivation of the land for this and the preceding crops, very few weeds will likely appear; but if thistles, docks, wild mustard, or other weeds which can be easily pulled, do appear, they should be removed before the flax is more than four inches high, as the trampling of the weeders on the crop always injures it to some extent, unless in its early stages of growth."

As to the time of sowing, Mr. Weir recommends " any time during the month of April, or early in

May." This is another form of the same error which put flax after every other crop in all respects. I am aware that in the system of husbandry prevalent in this country, the time mentioned by Mr. Weir is that allotted to flax growing, but it is TOO LATE in ordinary seasons, and for most parts of Ireland. In Irish husbandry as it has been, and is at present, I know of no part of the country, except a few places in the county of Down, parts of Antrim, Armagh, Tyrone, and Derry, where flax is generally sowed early enough. But I do expect such reforms in husbandry ere long (a result of flax culture) as will lead to the sowing of Flax at the proper time, which is for all parts of the country, except very late places, from Lady Day (25th of March) to the 20th of April. To sow earlier is less likely to be wrong than to sow later; but so much depends on the season, that I believe if the time above named be adopted as a rule, there will be sowing from the first of March in all early springs, and, except in late seasons, it will all be over before May-day.

I have been careful in this matter, as in all others, to recommend such reforms only as can be easily and profitably carried out. But in Mr. Weir's system there is a perpetuation of the old customs of the past history of Flax culture, and although it is not to be rejected merely on that account, yet I see no reason why new and better modes of culture should not be substituted, so far as they are soundly scientific.

In regulating the quantity of seed to the acre, it is necessary to measure the acre correctly, and not to guess at it merely as some do. It is also requisite to measure the seed. The right way is to allot the field off into small portions, and to measure the seed according to such portions, so that the seed may be equally distributed over every part of the field. Sowing should be broadest.

In cold soils, where drainage is defective, or the ground rough, extra seed ought to be used.

Thick sowing is better than thin sowing in any and in all kinds of soils. The proper quantities of new pure seed is as follows.

	IRISH ACRE.	SCOTCH ACRE.	STATUTE ACRE.
	*Imperial Quarts.	Imperial Quarts.	Imperial Quarts.
Riga ...	120	94	74
Dutch ...	104	82	64
English Irish American	120	94	74

* These quantities may be asked for at the seed shop, in the following way, viz. : for an Irish acre, 3 bushels and 3 pecks:—for a Scotch acre, 3 bushels, less 1 forpit:—for a statute acre, 2 bushels, 1 peck, and 1 forpit of Riga, English, or Irish seed. Of Dutch seed the quantities would be 3 bushels, 1 peck, for an Irish acre:—2 bushels, 2 pecks and 1 forpit, for a Scotch acre:—and 2 bushels for a statute acre. *If for raising seed, from 1/6th to 1/8th less according to soil.*

The following is an answer to a question quoted from the "Irish Farmers' Gazette" of the 23rd of January, (1864.)

"Plough the land in dry weather, harrow well, roll, and sow the seed broadcast, in the middle of March, at the rate of 3¼ to 3½ bushels per Irish acre; bush-harrow, and roll."

Some of the most successful Flax growers use more than the quantity of seed I have mentioned, I do therefore believe that the quantity named above by the "Irish Farmers' Gazette" is insufficient for even the very best soils, unless the growing of a crop for seed merely be the object, which it does not appear to be in the case referred to by the Gazette. I have known several cases in which three pecks have been sown on a statute rood, and one bushel on an Irish rood, with the greatest success; and judging from all I can learn, better profits followed than if a lesser quantity had been used; at the same time I believe the quantities I have given are sufficient, and not too much, and I am sure they agree with what is used by most of the best Flax growers in Ireland.

When the seed is in the ground, it should be covered by a seed harrow and a light roller, care being taken to cover the seed to *one inch* depth, and not more.

The sowing of Grasses with flax seed is bad economy, and although the *Irish Farmer's Gazette* does not condemn it, yet it is by no means to be approved. The following question and answer appeared in that Journal recently :—

" Could you lay down land with clover and grass seeds along with flax, and top-dress following spring ?

" You can lay down the land to.grass with flax, and it is constantly done."

This answer is, to say the least of it, careless ; and, judging by observation in the best Flax growing districts, I consider it calculated to lead astray. If Flax be allowed the undivided occupation of the soil, it will pay sufficiently to render . the growing of other crops with it unnecessary, from a profit and loss point of view.

If, however, any crop be put into the ground with flax, carrots may be tried with the greatest safety, provided the soil be loose and deep. The carrots ought to be sowed in drills, so that the "Flax pullers" might step over the drills, and when the flax is removed the carrots should be hoed, cleaned and manured with liquid manure.

A crop of rape or winter vetches might be got in some specially early seasons, but it is difficult to overtake any two such crops without sacrificing one in some measure. Therefore it is, on the whole, better to be content with good Flax, than in trying to get grasses, carrots, rape, or vetches, the Flax crop should have to suffer in any, even the least degree. Turnips of the Norfolk Globe variety is, of all " stolen crops," the best financially, and probably not the least likely to be successful agriculturally.

PULLING.

The PULLING of Flax ought to be done by active, able-bodied labourers, and always as soon as it is ready, which may be determined by the following marks :—FLAX IS READY FOR PULLING WHEN THE SEEDS ARE BEGINNING TO ASSUME A PALE BROWN COLOUR, AND WHEN THE STALKS ARE YELLOWISH ABOUT TWO-THIRDS OF THEIR HEIGHT-FROM THE GROUND.

The sooner flax is pulled after these appearances are visible, the better ; at the same time the palish or pea green sappy look peculiar to growing Flax should be altogether displaced by the yellowish tinge, only apparent when it begins to ripen, before pulling is attempted.

To pull too soon makes a weak soft fibre, which will neither yield well to the Scutching Mill or the Heckle. But to allow it to ripen too much makes a coarse fibre, which seldom sells for a good price. In case the Seed is to be saved for Sowing, the crop ought to be allowed to grow brown-ripe, irrespective of the fibre. Growing for both Seed and Fibre has never been satisfactorily accomplished in this country, nor would it be desirable to attempt it. It is, however, a question yet to be decided, and which it is to be hoped will soon get a practical solution—Whether growing for Fibre or Seed is the more profitable?

As soon as the Flax is pulled it ought to be RIPPLED, and got into the steep with all despatch. The Rippling Machinery ought to be in the field and at work while the Flax is being pulled. There seems no question as to the entire propriety of Rippling, except with very small farmers (or growers of half acre patches) who may not be able to get so little Rippled without keeping the Flax out of the Steep too long.

Of the two evils, it is perhaps the lesser to lose the seed of Flax grown for fibre, than to risk the injury consequent on keeping it too long in on the ground after being pulled.

If there were no other objection to the scheme, unhappily recommended by some members of agricultural societies and landed proprietors, of trying a little patch of Flax as an experiment, the fact that Rippling cannot be properly done unless at too great a cost, when the quantity is very small, would of itself be enough to deter persons of sound judgment from growing Flax at all, unless in sufficient quantities to give the crop a fair trial. Besides I am aware from observation, confirmed by the most reliable statements of disinterested persons, that of all classes of farmers those holding from four to six acres, are the LEAST likely to make a profit on Flax. I could also easily guess as much from my knowledge of the nature of the Rippling process, which, to be done properly, must be done by such a machine as no very small holder can afford to buy.

A Rippling Machine is very simple. A good sort of a Ripple may be constructed by screwing a set of Iron Teeth into a block of wood. THE TEETH SHOULD BE OF HALF-INCH SQUARE RODS OF IRON, PLACED WITH THE ANGLES NEXT THE RIPPERS, THREE-SIXTEENTHS OF AN INCH ASUNDER AT THE BOTTOM, AND HALF AN INCH AT THE TOP, AND EIGHTEEN INCHES LON This kind of a Ripple allows of a good spring, and saves the Flax from breaking.

The points of the Teeth should begin to taper three inches from the top.

WATERING.

After rippling the Flax should be taken immediately to the water. The greatest care and attention ought to be given to this part of the work, and the following directions are to be attended to, viz. :

2. IF SPRING WATER BE USED IT OUGHT TO BE COLLECTED IN THE POOLS AS LONG A TIME BEFORE PUTTING THE FLAX INTO IT AS POSSIBLE, (not less than four to six weeks) WATER CONTAINING IRON OUGHT TO BE AVOIDED. IF RIVER WATER BE USED, IT WILL DO TO LET IT INTO THE POOL THE DAY BEFORE THE FLAX IS STEEPED. The BEST SIZE OF A STEEP POOL IS 12 to 18 FEET BROAD, AND $3\frac{1}{2}$ to 4

FEET DEEP. PLACE THE FLAX IN LOOSELY IN ONE LAYER, SLOPED, WITH THE ROOTS DOWNWARDS, AND AT SUCH AN ANGLE THAT THE TIES OF ONE ROW WILL REACH THE ROOTS OF THE PREVIOUS ONE. The best covering is old lea sods, or moss sods, cut thin and laid close, and with a layer of Rushes or Ragweeds between the Flax and the Sods.

If rushes or ragweeds cannot be obtained, straw will do, and if sods cannot be got, stones will answer, provided the flax is sunk just so as to be under the water, and no deeper.

As the fermentation proceeds the weight of covering must be increased, but so soon as fermentation subsides the additional weight should be removed. Flax should never be sunk to the bottom of the pool. A small stream of Water ought to be allowed to run through the pool, or, if there are a series of pools in a line, the stream should be conducted along the side, and allowed to run into each pool separately, and the water of each pool run off on the opposite side.

ON THE AVERAGE FROM 8 TO 12 DAYS, SOMETIMES 14 DAYS, IS SUFFICIENT TO STEEP FLAX. Great care ought to taken that neither too much nor too little of the water be given it.

The following are the tests to try when it is steeped enough. TRY SOME STALKS OF AVERAGE THICKNESS, BY BREAKING THE SHOVE IN TWO PLACES, 6 INCHES APART, AT THE MIDDLE OF THE STALK, IF THE BROKEN BIT OF SHOVE (OR WOODY PART) COMES

OUT FREELY DOWNWARDS, WITHOUT BREAKING OR TEARING THE FIBRE, AND WITHOUT ANY FIBRE ADHERING, IT IS READY TO BE TAKEN OUT. This trial ought to be made twice (at least) daily. The change is sometimes so rapid after fermentation subsides, as to be very dangerous; therefore, for the same reason that a skilful herd will watch his sheep attentively at the lambing time, a good Flax grower ought to pay the strictest attention to his flax when in the latter stages of the steeping process.

Steeped or Retted Flax should never be roughly handled, or forked out of the pools. The most careful treatment will be amply repaid, however expensive.

Flax should be allowed to drain from 12 to 24 hours before spreading it, either on its roots ends, or in some way that will neither cause it to heat, or break or tease the fibre.

Many systems have been tried and some adopted, in substitution of the ordinary process of steeping in Pools or Rivers, and these modes seem on paper to possess some merit, and some of them look like decided improvements, but for some reason or other the old system has been but little superseded for so far.

I have thought it better to make no comment on the respective modes of operation in Ireland, for extracting the fibre from the plant, as I know that Flax growers will this season be asked to try not

C

a few plans never before heard of; and I believe the quantity likely to be grown will, at least, admit of all plans possessing merit being fairly tried. Flax is, however, too precious to be experimented upon, and therefore it is to be hoped growers will act with caution, and be slow to adopt changes from a system so long and so successfully carried out. I do not, however, believe that steeping in pools is the only mode that can be safely adopted; but I am quite certain it is 'better to keep to the old system, until a substitute in all respects worthy of adoption be not only found out, but also, by sufficient tests, fully shown to be worthy of confidence.

3. A Flax company was announced last season, from whose prospectus I make the extract below. If the proposals thus made could be carried out, it seems that a decided improvement would be introduced; but I am not aware as yet of the results of tests actually made, if any, nor as to the feasibility of the plans proposed, although the objects aimed at are unquestionably the very *desiderata* Flax growers ought to seek for. The company from whose prospectus I quote, was originally intended for England and Wales.

"The object of this company is to purchase flax from the farmers, and to afford increased facilities for the preparation of the plant in the various stages through which it has to pass from the grower to the spinner, and thereby to extend the home growth of flax, so essentially connected with a valuable branch of our national industry.

"The many difficulties attendant upon the extraction of the

fibre from the plant, have hitherto deterred landowners and farmers from encouraging its growth and preparation in sufficient quantities to supply the home market; although it is well known that English flax is peculiarly adapted to all descriptions of the linen manufacture. Out of about 150,000 tons of flax annually consumed in the United Kingdom, *not more than one-fourth is of native growth*, whilst hundreds of thousands of quarters of linseed are annually imported from abroad. This large importation involves an outlay of several millions sterling, the whole or a large portion of which would be saved, and added to the national wealth, by extending the home cultivation of flax.

"The long prevalent prejudice, that flax impoverishes the soil, has been dissipated by scientific and practical agriculturists, who have demonstrated that, if taken in proper rotation every eight or ten years, it does not exhaust the soil more than any other crop, and is as beneficial both for the farmer and landowners.

"The company proposes to establish *Rettories* in localities easily and cheaply accessible to the flax grower, and in which large breadths of the plant can be favorably cropped, where all the processes required for preparing the fibre from the raw plant will be carried on. These establishments will supply a want long felt by the grower, and render flax, as a crop, valuable in cereal farming.

"The several marketable products from the Rettories will be:—

"*Assorted Seed for Cropping.*

" *Conversion of the Inferior Linseed into Oil and Oil-cake.*

" *Prepared Fibre for Spinners from the best portion of the crop.*

" *Best Tow for Spinners.*

" *Refuse Tow, Shoves, and the inferior portion of the crop wrought into paper-pulp.*

" *Liquor or Drain from the steeping-vats a valuable basis for Manure.*

" *Chaff from the Boll suited for feeding Draft Horses, Milch Cows, &c.*"

If all this can be done in any other place, it can be done most profitably in Ireland than elsewhere.

SPREADING.

After Flax is watered, it has next to be spread on the ground to bleach. Clean, short, thick, pasture land should be selected for spread-ground. THE SWARD SHOULD BE MADE LEVEL, AND THE FLAX SHOULD THEN BE SPREAD EVENLY AND THINLY ON THE GRASS—IT OUGHT TO BE KEPT ON THE GRASS 6 TO 8 DAYS IN WET WEATHER, AND 8 TO 10 DAYS IN DRY WEATHER, AND TURNED TWICE OR THRICE WITH A ROD 8 FEET LONG, AND $1\frac{1}{2}$ INCHES DIAMETER.

When there is an appearance of rain Flax on the ground should be turned that it may be beaten down a little, which keeps it close to the ground, and hinders entangling, blowing away, and scattering. Flax is ready to be lifted when it is easily scutched; this may be tried by hand-beetling and scutching a little. Another way of knowing when it is grassed enough, is, that many stalks rise up in the field like a *bow and string.*

LIFTING.

LIFTING should be very carefully done, and DRYING also should be most minutely looked after. *Fire-drying is dangerous—sometimes destructive to the quality of the flax, and ought to be avoided.* Building the " *beats*" into small stacks, through which a current of air can pass, is a good plan ; but as so much depends on the weather and the class of out-building at a farmer's command, it is better to say merely in a general way that *flax ought to be dried thoroughly, but not on kilns or by fire.*

SCUTCHING.

FOR *hand-scutching, which is very objectionable,* it is necessary to beetle the flax carefully ; but as Rollers are now a part of every Scutching Mill machinery, home beetling has not to be done. Scutching is decidedly skilled labor, although any person may be taught it in a very short time. Too much importance cannot be attached to this part of the process. Farmers should either unite in erecting suitable Scutching Mills, or devote such a portion of their soil to FLAX crops regularly, as would encourage capitalists to get up Scutching Mills on the very best style, and with the newest and most approved machinery.

The want of Scutching Mill is not likely to be any longer felt when it is understood that a breadth of Flax sufficient to employ mills at convenient distances from each other will be sown. A mill of *three stocks* will scutch the produce of 50 statute acres in 70 to 80 working days of 10 hours each. Mills of 6, 9, and 12 stocks respectively, will scutch in the same proportions.

For the guidance of persons about to erect

Scutching mills, I wish to say there is greater economy in 3, 6, 9, 12, and 18 Stock mills, than 2, 4, 8, or any number that will not divide by 3. The reason is, one scutcher can "*buff*" as much as two others can "*clean*" and "*finish*." Besides, it is desirable that the same hand should buff, clean, or finish constantly;—except for these reasons however, it matters not, so far as it affects the Flax, whether a mill has 3 or 20 stocks. A well-constructed mill should scutch at the rate of 14 lb. per stock per hour. Each stock has from 4 to 8 handles.

Portable Scutching mills are now much employed, and where Flax is not largely cultivated they are very useful. Some of these which are no weightier than 15 cwt., have *two stocks and a set of rollers*, and can be attached to any kind of power, or can be worked by hand, and will Scutch about 8 stone per stock per day. The subject of Scutching mills, however, belongs not so much to the growing as to the dressing of Flax ; and as this department deserves a pamphlet to itself, I will forbear any further remarks, except to add, that the greatest care ought to be taken to have Flax properly scutched, as from 6d to 1s. per 14 lb. (or about the cost of Scutching,) of the price depends on the manner in which this part of the process is performed.

It is neither so much (if at all) the climate or soil as the manipulation of the crop, that causes Belgian and other fancy foreign Flax to so far exceed Irish, as to be worth double the price.

INDEX.

INDUSTRY IN IRELAND.

A TREATISE

ON THE

AGRICULTURAL POWERS, MANUFACTURING CAPABILITIES,
AND COMMERCIAL ADVANTAGES OF IRELAND.

BY

WILLIAM GLENNY CRORY,

AUTHOR OF "THE COMMERCIAL IMPORTANCE OF IRELAND'S INDUSTRIAL RESOURCES;"
SPECIAL COMMISSIONER OF "THE DUBLIN EVENING MAIL."

LONDON: SIMPKIN, MARSHALL, AND CO.
EDINBURGH: JOHN MENZIES.
DUBLIN: McGLASHAN AND GILL.

1865.

PREFACE.

THE development of Irish resources and the progress of Irish manufactures constitute a subject not possessed, in one sense, of the charm of novelty. These phrases have indeed been so familiar that a new writer is naturally apprehensive of being mistaken, and of seeing his book put aside with a shrug or a sneer. Irish manufacture movements, unfortunately, have proceeded upon false principles, and have failed for that reason; and were the following pages merely devoted to the old rhapsodies upon this theme, they would deserve to perish. At the start, however, the author wishes distinctly to state that he is connected with no "movement" of any kind, intended to force any Irish manufacture, or Irish resource, into a factitious prominence, but has felt it incumbent upon him to put before the public, in this form, his experience of many years upon Irish manufacturing capabilities, and the profit which they offer to Irish and English capitalists. He has striven to look at the matters that have come before him in a purely business light, and disavows all unmercantile enthusiasm. Having resided for a long period in England, and being familiar with the modes of thought of moneyed men on both sides of the Channel, he flatters himself that he never forgets the spirit which must prevail in all practical enterprises, if they are to be successful. Having taken up a position of this kind, the author may probably be allowed further to say that his book is, nevertheless, a novelty, in being literally the first attempt which has been made in a momentous manufacturing crisis, to point out, in detail, where Irish resources lie, the conditions that affect each description of them, and the prospects of return, in special places and departments, for the investment of money. The writer is, of course, not insensible to the noble efforts of those who have preceded him in this walk, with a special title to interest the public, and, in particular, need hardly mention Sir Robert Kane's book; but he claims to have the peculiarity of fastening the attention of the practical men of the empire upon Ireland as a field for enterprise, at an epoch which, in his mind, is emphatically her " opportunity," and one that, if neglected, may not speedily recur. The convulsion we witness in Lancashire by the " cotton crisis," which, in the opinion of the most far-seeing, will not subside with the cessation of the war, has established the imperative necessity for cheaper production and lesser freightage, and the writer has undertaken to prove that these objects can only be attained by an immediate and extensive utilization of the water-powers of Ireland. He has also shown that no difficulty of obtaining skilled factory hands, familiar with Irish manners, and prepared for a contented residence in this country, presents an obstacle to the working of Irish mills, since late experience in Lancashire has elicited the fact (well known before to all practical men) that a large number of the operatives are Irish, and of Irish extraction, to whom nothing would be more congenial than a return, under comfortable circumstances, to their native land. An identical facility exists, moreover, for obtaining skilled hands to work in Ireland, in the manufacture of hardware, cutlery, machine-making, china, glass, and chemicals, which make up the cargoes "outwards" of ships trading in cottons and foreign and colonial produce.

The author also devotes considerable space to the topic of flax cultivation, for which Irish soils and the Irish climate are specially fitted. He was the advocate of a vastly increased growth of this plant in the DUBLIN EVENING MAIL (the journal in which his letters first appeared), at a time when the public had not yet begun to recognise the importance of the occasion. Since then he has had the satisfaction of finding his views gaining ground with thoughtful men; and, while he would guard himself against helping to create a pernicious flax mania, he is convinced that when flax-growing and flax-spinning in Ireland have attained their legitimate position, the opportunity for cotton-manufacturing in this country will also be more clearly apparent, and more valuable.

The beneficent social effects which would infallibly result from any general introduction of manufactures into Ireland—the breaking down of prejudices, the uniting of classes, the propagation of sounder principles, and the relieving of agriculture from the strain of unemployed labour—would be an achievement so incalculably valuable to the State and the People, that the philosopher, the philanthropist, and the statesman must equally feel interested in the discussion of the subject, when pursued in no visionary way, and associated with no declamation, but treated as a serious matter of figures and of facts.

The author, therefore, encouraged by knowing that the perusal of his letters as they appeared daily in the public press, has already produced fruit, publishes them in a collected form, trusting that they will be found in this shape more suggestive, more consistent as a whole, and more likely to bring about the desired result—the investment of capital in enterprise in Ireland, at a period when the manufacturer-princes, discerning difficulties ahead as well as actually existing, are casting about them for fresh instruments and new fields of effort.

INTRODUCTION.

I BELIEVE the only reason I have for writing an introduction to the following pages is a sufficient one, as, without some explanations which I could not otherwise give, my book might be misunderstood. My desire is merely to bring the neglected resources of this country most prominently under the notice of my fellow-countrymen at home, to introduce them more fully to the British public, and to tell Irishmen who have made money in foreign lands, or those who hope to return to their native country, that there is a field for the investment of capital and the employment of labour in Ireland, obviously not to be excelled, and probably not to be equalled elsewhere on the globe. I have endeavoured, in the discharge of a duty so onerous, and in doing a work so formidable, to treat every material noticed in the following treatise, so as to leave the impression on my readers' minds that I only mention their existence, and hint at some points of their worth, but do not at all even attempt to describe minutely, or to value fully, resources in whose description volumes might be filled, and to express the full value of which I do not know of sufficient figures. I have also dealt with each of these as if it were a witness by whose testimony capitalists, manufacturers, and merchants would be guided in the investment of money, the exercise of talent and industry, and the operations of commerce; but I trust geologists, mineralogists, and others more competent than I am to appraise the value of Ireland's industrial wealth, may be induced to test the accuracy of my statements by a cross-examination of these witnesses in the cause of Ireland's improvement. I have written with the knowledge clearly before my mind that my subject is at once unpopular and important. I have not overlooked the probability that the pages I now submit to public notice may be read by persons in this country who make it their business, whenever an effort is made to bring the subject of Ireland's resources of industry before the people, to offer objections to the outlay of money in their development, which is mostly done by recalling the failure of some ill-advised or badly-managed experiment, or by exaggerating the few slight difficulties which must naturally present themselves in the earlier stages of Ireland's progress in manufactures, as a consequence of England's having already gained a position in like undertakings. I have, therefore, given the prominence to such facts as I thought would enable any one desirous to deal fairly with the subject, to meet such fallacious notions as are founded on ignorance, but which, because of being often propagated by persons supposed to be well-informed, have, for so far, paralysed the efforts of some, and deterred or alarmed others,—whilst they were also no less powerful in leading strangers astray than in stopping industrial efforts at home.

I have also the hope that my book may be perused by many out of this country, amongst whom, doubtless, there will be some who may look with jealousy at the position I assume for Ireland in manufactures, to whom I trust I have, in the small space at my disposal, offered sufficient reasons for regarding Ireland's elevation to the first rank in Europe as a commercial country the *desideratum* of the further expansion in British commerce. On the development, therefore, of Ireland's resources depend issues which render the doing so immediately much less a national than an Imperial question.

With a view to enable my readers to understand more easily some ideas which are purposely implied, but not pointedly stated, I take leave to announce that I have written with a desire to accelerate a general tendency which I observe with inexpressible pleasure in all the movements of commerce, namely, that all mercantile operations, to be successful, must be guided by enlightened minds, and conducted under the most rigid adherence to the self-same principles of generosity and benevolence as are expected in other walks of life. Besides, recent events in these and other countries show that before they can be profitable, the interchanges of commodities, and the financial operations of merchant with merchant, or of the merchants of one country with those of another, must be guided by laws as carefully defined as those necessary in the laboratory of the chemist, or the workshop of the mechanic.

I have written, moreover, believing the time is not far distant when all will admit it to be as necessary to have a special department in our schools and colleges for the teaching of the science of commerce, and the proper preparation of merchants for their profession, as it is now to make provision for the proper qualification of Christian teachers, lawyers, surgeons, doctors, or those of any other profession. Obviously, more of the interests of a country and the progress of civilization, in the widest and highest sense of these terms, depends on the mode of conducting commercial operations than can possibly attach to the manner in which persons in any other position of life discharge their professonal duties. Uneducated merchants can have no place in the future of commerce, nor can avarice any longer be safely taken as a guide in getting rich, much less can the disregard of any material or discovery wherever to be had, or from wheresoever emanating, be longer tolerated; but, on the contrary, mercantile operations, to be successful, must be governed by moral rectitude, and not as now, to a large extent, by an unenlightened eagerness of gain. I saw whilst com-

posing matter for these pages (what I see now more than ever)
the necessity of casting off the repulsive dictums of (so-called)
political economists, whose crude notions about supply and
demand, and cold heartlessness in reference to the relations of
capital and labour, have begun to bear such bitter fruits as the
" cotton crisis " in England, by which half a million of people
would have been reduced to starvation these last two years
but for the benevolence of the public in all parts of the world
so nobly manifested, and in which Ireland has had an honorable
share; as also in the discouragement of industry in Ireland, by
which probably a whole million have been kept in deep priva-
tion these last ten years, and who would have been " swept off"
but for aid afforded time after time, in which England always
cordially joined.

Commercial men must reject all sentiments unsupported by
sound reasoning, and avoid every policy incompatible with the
constant observance in the strictest manner of the precepts
which demand from man to man the offices of brotherhood, and
the expressions of love. But I must not fail to take advantage
of this opportunity of expressing indebtedness to the kindness
of those who gave me information, or otherwise aided me whilst
the substance of my work was passing through the columns
of the " Dublin Evening Mail," or under revision for publication in a book form. For several suggestions, I am indebted
to the Rev. R. B. Kidd, LL.B., Belfast; as also for proposals of
further assistance, which I trust I may have the pleasure of
availing myself of in a future edition. I have also to acknowledge valuable information on flax and cotton manufacture
received from P. Macauley, Esq., J.P., Cleggan, Co. Galway;
and James Harden, Esq., Chapelizod; and in regard to
machinery used in flax mills I received several valuable facts
from J. B. Mitchell, Esq., engineer, Dublin. The patentee of
that invention to which I refer in several places (James S.
M'Ardle, Esq.), not only gave me valuable assistance in re-

ference to seaweeds, but has furnished me with facts on flax manure, which are too valuable to remain unknown, and which I hope to use in some way in fulfilment of Mr. M'Ardle's desire for the good of the people of this country. From John Kelly, Esq., of the Museum of Irish Industry, I also got much guidance in the treatment of the subject of stones and marbles. To none, however, am I more indebted than to the proprietors of the " Dublin Evening Mail," and very especially also to James Scott, Esq., well known in connection with that journal and the " Warder," to whose kindness I am under obligation for incalculably valuable assistance, well-timed suggestions, and much encouragement in my labours on the work for which, with all its defects, I now ask the consideration of all who are interested in commerce, or who desire to see Ireland fill the place in fact, for which she is fitted by nature.

W. G. C.

Dublin, 1st January, 1863.

CONTENTS.

APPENDIX.

A TREATISE,

&c., &c.

----◆----

CHAPTER I.

INTRODUCTORY.

IRELAND has already been the theme of so many speeches and
the subject of so much writing, that it seems, at first sight,
impossible to place either her present state or future hopes in
any new light; but this apparent difficulty vanishes before a
close examination of the neglected resources of industry in the
country. A very cursory consideration of the latent *materiel*
for manufactures and commerce will, probably, convince those
most doubtful of Ireland's capabilities that, according to all precedent in cases of like resources for the employment of labour
and the investment of capital, she may be placed first on the list
for immigrants.

In dealing with this subject, we will set forth the manufacturing capabilities of the country in detail. We shall describe
their character, and point out their locality, so that capitalists
may know to what points their inquiry ought to be directed.

In no aspect does the importance of a fair exposition of
Ireland's industrial capabilities seem so important as in its
relation to emigration. Hitherto the people of Ireland, and
especially those of the midland and south-western parts, have
rushed to every country on the globe, according to their means,
in search of employment, materials for which, better than they
can find anywhere else, are neglected at home. This emigration, however, though fraught with many bad consequences,
was not wholly an evil. Yet such is the point it has already

B

reached, and such are the concomitants of its operation, that to allow it to go further would be most culpable negligence, whilst to excite or encourage it is assuredly to inflict the greatest possible injury upon the interests of both British and Irish commerce.

That division of families, consequent on emigration in the past, may not only lead to some further emigration, but may also become the occasion of many returning to this country. This will be the more or less visible according as the country improves more or less rapidly. We are satisfied that to enlighten the minds of people everywhere on the subject of this country's resources, would become the occasion of bringing thousands of remigrants and hundreds of thousands of pounds sterling back into Ireland.

We do not expect that opinion should be considered evidence in this case. The only testimony we attempt to bear is such as facts, too plainly fixed before common observation to be contradicted, will speak. We find these rise up before the inquirer after manufacturing capabilities in this country in the following order, in which order also they will be treated in these pages.

Firstly, water-power, cheap rents for mill-sites, comparatively cheap and very abundant labour, suitable for cotton manufactures and woollen manufactures, both of which might be pursued with the advantage of that division of labour, by which operations in all departments are reduced to a minimum cost. Also, those engaged in the linen trade might avail themselves of the cheapness of motive-power, and the superiority of water-power to steam-power in the spinning, but especially the weaving of linens. Bleaching may be done as well in Mayo, Clare, or Galway, as in Down or Antrim.

But we do not ask anyone to consider the claims of Ireland's resources of industry merely because their development would benefit Irish people, but rather expect their development to take place because it would benefit those who engage it. We expect people from England and Scotland to go to Connaught and Munster, and people from Ulster to go westwards, where obviously this development must begin, if begun aright, simply because any given sum of money laid out in the trades named above would be so productive of profit as to form a sufficient inducement for the most rigid economist to choose Ireland.

The question is not now, as formerly, Are there resources of industry in Ireland sufficient to support her people? This has long ago been answered incontrovertibly in the affirmative. But it may be asked, Do these resources exist to such an extent, are they distributed in such a manner as to make their utilization practicable at the present moment, and are these resources so valuable as to induce capitalists to turn their attention to their development for sake of differential profits likely to arise from the employment of enterprise in this new field, rather than in the older channels of trade?

On the single point of superior manufacturing capabilities these questions turn. The maritime advantages, the geographical position, the nearness of any part of the country to the seaboard, and of the island to the markets of the world, in combination with the differential profits to be set forth, will prove that to the West of Ireland much manufacturing enterprise from Ulster and from Great Britain may be safely transferred.

Fisheries must still be enumerated amongst neglected resources. Home markets for farm and garden produce will also be noticed in their relation to the interests of agriculture. Marine plants are also a much-neglected resource of industry.

We shall most specially notice peat bog, also glass and earthenware manufactures, the extension and value of which are highly important, especially as a means of supplying back freights for ships carrying raw cotton, wool, flax, hemp, jute, and other imports necessary in trades and manufactures.

We will also notice the existence and value of granite, marble, and other building materials. Banking will also be dealt with.

The following table, which sets forth the estimated relative cost of working cotton mills in Ireland by water-power and steam-power respectively, places the case in a form which we believe may be understood by any one, and is likely to be plain to those conversant with cotton manufactures :—

Estimated Cost of Water-power for Cotton Manufacturing on or near the Seaboard in Ireland, as compared with Steam-power in Manchester, or any other part of England, where Raw Material and Manufactured Goods are subject to a Land Carriage of, on the average, 35 Miles, and which would be either wholly, or to a great extent, saved in Ireland, besides being at a Port for Shipping six to eight Days nearer any Foreign Market.

Per annum per 100 (actual) Horse-power—500 hands.	Water-power in Ireland. £ s. d.	Steam-power in England. £ s. d.	Difference in favour of Ireland. £ s. d.	Difference in favour of England. £ s. d.	Rate per cent.
Rent In the proportions as £8,000 is to £4,000, at 8 per cent.	240 0 0	320 0 0	80 0 0	...	25
Taxes	60 0 0	100 0 0	40 0 0	...	40
Interest on cost of chimney, engine and boiler, and second boiler (£2,100)	...	105 0 0	105 0 0	...	
Do. on cost of 80-horse water-wheel and fixings	50 0 0	...	...	50 0 0	
Wear and tear on engine and boiler, 10 per cent.	...	210 0 0	210 0 0	...	
Wear and tear on water-wheel, 5 per cent.	50 0 0	...	...	50 0 0	
Cost of machinery for and including, say 20,000 spindles and 600 looms, about £18,000 to £15,000	...	...	...	...	
Fuel, 218 tons 16 cwt. 3 qrs., at 7s. 6d. per ton	...	80 8 9	80 8 9	...	8
Water for boiler	...	25 0 0	25 0 0	...	
Engineer's wages, oil, tallow, &c.	...	106 0 0	106 0 0	...	
Water-rent, £80; fuel for keeping up temperature, £40; man attending sluices, £80	150 0 0	...	...	150 0 0	
Fire insurance	30 0 0	50 0 0	20 0 0	...	
Wages in the proportion on the average as 9s. per week in England is to 8s. in Ireland	10,000 0 0	11,250 0 0	1,250 0 0	...	
	10,580 0 0	12,246 3 9	1,916 8 9	250 0 0	Or
Difference in favour of Ireland	1,666 3 9	...	...	1,666 3 9	18 6-10 per cent.
	£12,246 3 9	12,246 3 9	1,016 8 9	1,916 8 0	

In calculating the above table, we have fixed the rent according to the cost of erecting buildings in the West of Ireland and the cotton districts of England—say Galway and Manchester, respectively. A stone building, with inside brickwork, in the former, would pay an owner at six per cent. as well as outside brickwork would pay in the latter at eight per cent.

We have been careful to keep the outlay for chimney, engine, and boilers as low as possible, whilst the sum set down for a water-wheel is the very outside of that which would be necessary in the average of cases, even if the least eligible sites were chosen; but eligible sites are so numerous in Connaught and Munster that none except those having the best waterfalls need be chosen, even under the most rapid extension possible, during the next fifty years.

The Turbine water-wheel is not known yet to that extent in this country which would justify any expectation from the savings by its adoption in the first outlay of money for creating motive power. But it is to be hoped engineers will give this class of wheel more attention.

The adoption of the latest inventions might add some few thousands to the cost of machinery, but would lead to lasting economy. It would also illustrate that great principle not yet fairly brought out in the cotton trade, namely—that the less manual, and the more machine labour, the better for all interested in the trade, especially for the operatives.

In the amount of wages put down for Ireland and England there is no attempt at fixing the average cost per day for any given class of mill, but merely the relative cost in the respective countries. This proportion is founded on a permanent basis, not to be moved, supposing Ireland to become at once—as we hesitate not to say will yet be the case—as celebrated for flax-growing as India and the Southern States of North America are now for the culture of cotton.

Ireland could be made to produce profitably to the farmer seven times as much flax as is now raised, and might thereby become the means of enabling flax-spinners to expand a trade which, though it has grown much since 1851, is never likely to assume its natural proportions till Ireland's broad acres, to the extent of one million at least per annum, are covered with flax. To raise the yield to this extent would give occasion for the introduction of at least 100 or 150 more spinning mills, whose produce would necessitate the use of power-looms in every branch. Even in case of such extension, food would be still cheaper in Ireland than in England; for the advantage of nearness to the corn-growing and cattle-feeding States of America, and, in fact, to every country in the world whose trade is worth notice, would be greatly in favour of cheaper living in Ireland, consequently of lower wages. In fish alone there would be a large saving; house-rents are also cheaper. The difference in wages, being therefore founded on an abiding basis, is never likely to be disturbed by any possible changes.

The primary object of the above table is to put a fact before the public in a forcible form, namely—that cotton manufactures may be carried on in Ireland at a profit 13 and 6-10ths per cent. larger on 1-3rd the whole " turn over" of the trade than can be realized in Great Britain.

The secondary object is no less important, namely—to show capitalists, as such, that money may be invested in the cotton trade in Ireland (as will be more fully explained afterwards), so as to yield as large a profit to the manufacturer as he can make in England, after paying the capitalist a liberal percentage for the use of money.

Water-power, eligible sites for mills, as well as some buildings already erected, with land for workers' houses and gardens, can now be had at either the ports of Limerick or Galway, or at easy distances by water carriage, to the number of 625 such mills, which could even be increased so as to work an equal quantity to that now employed in the United Kingdom in linen, cotton, and woollen, and in a district whose seaboard would extend from Waterford to Westport, both places and all intermediate included. It is also a fact, that in every manufacturing town in the United Kingdom, and in America (e. g., Lowell, Massachusetts), a very large number of the operatives are Irish, and most of them from Galway, Mayo, Tipperary, Roscommon, King's and Queen's Counties, Limerick, Clare, and Waterford. It is also obviously true that in Connemara, and other parts of Galway, in Mayo, and in fact in several Irish counties, hundreds of hands trained in Manchester, Glasgow, Belfast, or other places, are now residing, most of them comparatively idle, and all of them willing to earn money, especially at what they best understand—namely, flax, or cotton spinning and weaving. Every mill employing 500 operatives would, directly and indirectly, give employment and support to at least 2,000 persons. Besides this, the erection, fitting up, and working, would call into profitable use at least £25,000 over and above those sums necessarily circulated indirectly in shipping, &c.

CHAPTER II.

COTTON MANUFACTURES—DIFFERENTIAL PROFITS.

The table in Chapter I. was intended to set forth, in the simplest possible form, the outlay for rent, motive power, and wages of a moderate-sized cotton factory, as worked respectively by steam-power in England and by water-power in Ireland. But as steam-power and water-power are both suited for a cotton mill, the custom of the trade is to adopt in any given case whichever is more available, the preference is, however, given to water-power, since it is not only cheaper, but more steady and manageable.

The sum necessary to erect a mill of the size named, to put the best machinery into it, and to leave a surplus, at the same time, for working capital, may be estimated at £25,000 to £30,000; and, consequently, the saving of £1,666 3s. 9d., shown in the table contained in the former Chapter, would be at the rate of 5 11-20ths per cent on the larger, and 6 7-10ths on the smaller of these sums, respectively, per annum.

In ordinary cases, the lesser of the above-mentioned sums would be amply sufficient; but the larger amount might possibly be employed advantageously in adopting the very newest inventions in machinery, and especially those by which the greatest possible amount of work could be done with the fewest possible number of unskilled labourers. In either case, however, the saving would be alike valuable and important.

But every one who is otherwise qualified for going into the cotton trade in Ireland, and would be inclined to do so for the sake of the special advantages it offers, cannot be expected to possess the whole capital necessary for so large an undertaking as we have spoken of; in which case the *differential profit* becomes still more valuable for practical purposes, as it is sufficient to pay for the use of money, and *still to leave the ordinary profit of the manufacture intact.* Even supposing the extreme case of a manufacturer borrowing the whole of his capital at 5 per cent., he would, after paying interest, be as well off in Ireland *as he would be in Great Britain with no interest to pay.*

Another aspect of importance in the saving set forth in the table appended to Chapter I. is, that it suggests a sufficient reason for persons of limited means borrowing a part of the capital necessary for the trade ; whilst, at the same time, it (this "differential profit") constitutes a sort of collateral security to the lender, whether on mortgage or personal security, for the repayment of principal and the regular defraying of interest.

To a certain class of capitalists the very fact that money would be required for the expansion of the cotton trade in Ireland upon a sufficiently large scale, must be a cheering consideration. We refer to those hundreds of persons who have only such a sum of money to live upon as barely affords them a support at the ordinary rate of interest, but is not enough to carry on business of a class compatible with their position in society. The cotton trade in Ireland would afford an opportunity for such persons lending their money, to be repaid in sums extending over ten, twenty, or a greater number of years, and to bear interest at 4, 5, or 6 per cent., as might be agreed on. Besides, there are in this country, in England, and in Scotland, persons trained in mills, who have also great practical experience and some capital, and who would gladly invest what they possess, provided they could get as much in addition, to be repaid by instalments out of profits, as would enable them to carry on trade effectively. There are also many capitalists whose object is not to leave money after them, but to live upon their money comfortably. The cotton trade as it is, and the linen trade, if expanded, seem to afford the connecting link between these parties.

Ireland's nearness to the market for raw material may be estimated at a given sum, and added to the £1,663 3s. 9d., "differential profit," already arrived at. The saving on freight of raw cotton for a mill of the size assumed would be, say 600,000lbs., which, at 1-32 per lb., (allowing the quantity requisite for a fair variety of production) would be £78 2s. 6d., making, together with the above sum, £1,744 5s. 3d. in favour of Ireland, or say, in round numbers, £1,740, which would be available to pay interest on capital, without touching ordinary profits.

The table No. 2, at foot, illustrates the meaning more fully. We ought to say that this plan of raising capital has been tried with most marked advantage to both borrowers and lenders in hundreds of cases in England. Its consideration, therefore, may be strongly commended to merchants, shopkeepers, and others who have small sums of money to put to use for a time, which they would not like to fix on mortgage for a period of years.

Capital is seldom wanting where there is a fair prospect of good interest and ample security. In this case the security would be unquestionable, and a fair rate of interest may be easily paid. We believe if people but begin to consider the subject of Ireland's water-powers, that hundreds of thousands sterling, now latent, would soon be brought into profitable use. It is a very remarkable fact, and one altogether apposite, that joint stock companies, even since the Limited Liabilities Act came into operation, do not grow in Ireland as might have been expected. One would have supposed that the undeveloped capabilities of the country would have led to the formation of many such companies; but it is not so. This may be accounted for by the prevalence of that notion which led the Irish people, in particular, to look to others to do for them what they ought to have done themselves. Perhaps, more frequently, the reason is that those who are the best fitted to estimate the value of the country's resources are, of all others, the most ignorant as to their existence; or, it may be, the indiscriminating criticisms of some persons on Joint Stock Companies, because of their abuse, have deterred the people of this country from availing themselves of the Limited Liability Act. But as English capitalists have actually already offered to take mills in Ireland, if built in suitable places, surely it would pay better to lay out money in building cotton mills, which would readily let for six per cent. to eight per cent. on the outlay, than to keep money idle, or to invest it, as many do, at less than half that rate of interest. The building of cotton factories would not only raise portions of ground now useless to a building ground price, but would also enhance the value of every acre contiguous to any such erections.

Next to these considerations the position of the towns and harbours in the South and West of Ireland, and especially the port of Galway, in relation to America, offers facilities for trading to an extent equalled in no other part of Great Britain. The growth of manufactures always leads to the increase of shipping, and this increase, operating on Galway, would raise her to that place, in fact, which she now occupies by nature—namely, the first mart for American produce in the ports of Europe.

If cargoes of cotton had to be divided with the Irish cotton-spinners to a large extent, it would be more business-like and greatly less expensive to tranship what might be wanted in Liverpool by steamers from alongside ships in the port of Galway, Cork, or even Limerick, than it is to do, as is now done in Liverpool, with corn, flour, and provisions which find their way to Ireland. The advantages to the Irish consumer of such a reformed mode of proceeding would be no less important than the adoption of the common-sense mode of action would be useful to British commerce. English people, though they may have less than Irish people to gain by the extension of cotton manufactures to Ireland, have nothing to lose by such a course. The present falling off in the supply of cotton is, as an admonition to prepare for the future, rendered still more significant to the

British manufacturer by the rapid progress of the French and other foreigners in producing that variety of goods which, as a result of his cheaper production, enabled the British merchant hitherto to hold the first place in the markets of the world. Cheaper production is impracticable except by the substitution of water-power for steam.

The table shows how the differential profit could be applied in paying off principal and interest on fixed and floating capital, say £25,000 :—

	Interest at 5 per cent.			Principal.					
	£	s.	d	£	s.	d.	£	s.	d.
1st year	1,250	0	0	490	0	0	1,740	0	0
2nd „	1,225	10	0	514	10	0	„		
3rd „	1,199	15	6	540	4	6	„		
4th „	1,172	4	8	567	15	9	„		
5th „	1,143	16	0	596	8	6	„		
6th „	1,114	0	8	625	19	9	„		
7th „	1,082	14	8	657	5	9	„		
8th „	1,049	17	0	690	3	0	„		
9th „	1,015	6	0	724	18	8	„		
10th „	976	2	8	763	17	9	„		
11th „	940	18	8	799	1	9	„		
12th „	859	0	8	890	19	9	„		
13th „	814	9	0	925	11	0	„		
14th „	768	8	0	971	16	8	„		
15th „	719	1	9	1,020	18	8	„		
16th „	625	11	0	1,087	9	0	„		
17th „	598	13	6	1,141	6	6	„		
18th „	541	2	8	1,198	17	9	„		
19th „	481	3	8	1,258	16	9	„		
20th „	418	4	6	1,321	5	6	„		
21st „	352	2	8	1,387	17	9	„		
22nd „	282	14	8	1,457	5	9	„		
23rd „	209	17	0	1,530	3	0	„		
24th „	133	7	0	1,606	13	0	„		
25th „	53	1	0	1,008	7	0	1,061	8	0

CHAPTER III.

THE COTTON MANUFACTURES—DIFFERENTIAL PROFITS (CONTINUED).

No one who has looked attentively into the condition of the people of Ireland, could have failed to see a marked difference in favour of the inhabitants of Ulster, as compared with those of the other provinces; yet, if the positions to which Irishmen have attained out of their own country were taken as the chief test of merit, that distinction in favour of Ulster would become much less clearly defined. This fact becomes relevant to the case before us, because (especially in the South and West) too much is expected from Ulster people, and too little reliance placed upon those of other parts, in regard to industrial progress. It is true, nevertheless, that Northerns, in large numbers, have done as well at home as abroad; whereas, such can scarcely be said truly of Westerns or Southerns, except in a comparatively few cases.

The natural influence of self-reliance is to stir people into activity, and fix them down to the patient prosecution of practical enterprises. It is not three centuries since Ulster was a morass: it is now a garden. But negligence has, in even a shorter period, allowed other parts of Ireland, naturally far richer, to become a comparative wilderness.

The admission of the principle that industry is more important than rich resources, is a necessary preliminary to every other step towards manufacturing success. A differential profit of £1,740 per annum, on a mill employing 500 operatives, is certainly very great, but even this may prove insufficient if the people who would be benefited by the introduction of manufactures do not hail the fact with a hearty welcome. In placing industry before natural resources, we by no means assume that the resources of all places are alike valuable. The very contrary is forcibly illustrated in the greater success than was possible at home of many Ulster people in more favoured parts of the country. That plodding, careful, constant attention to business which has told so favourably in the North would, if applied, tell just so much more favourably in Connaught and Munster, as the undeveloped resources of these are superior to those of Ulster. The chief reason why we urge the West on the people of the North is, simply that we wish the great industry of those in that part to find the best field for its employment. That field is Connaught and Munster, and not Canada, or Australia, or New Zealand.

· We are happy to find that Ulster people have already begun to go Westward, and that the numbers of these laudable adventurers are rapidly increasing.

But the fact that many who are indolent at home, work hard out of Ireland, has often been mentioned as a reason why success cannot reasonably be expected to follow manufacturing movements dependent on Irish labourers. But we believe that this fact not only gives no handle for such fallacious excuses for further neglect, but, on the contrary, merely shows that Irishmen, like all others, will only work when fairly rewarded for their labour. It ought rather to urge upon all who wish to enlist Irishmen in industry, the necessity of paying them fairly for their work. If there were as good wages, there would be as hard workers in Ireland as in England or America. Irishmen know, moreover, that 20s. in Ireland is as good as 22s. 6d. in England, or 24s. elsewhere.

After the closest observation in this country and Great Britain, and the most careful comparison of what we have seen, it is our opinion that the lower the current rate of wages in any particular district or country, the more expensive, on the average, is the cost of labour. On the other hand, the higher the rate of wages, so long as the labour market is fairly supplied, the cheaper is labour.

Cheap labour, in the good wages sense, is always well executed; dear labour, in the low wages sense, is never rightly performed.

Obviously, Ireland, in respect to manufacturing industry, is at present in something like that state in which the ocean is after the breaking up of ice. Ignorance still floats about, though it is daily lessening before the influence of information. Prejudice still exists, though it is weakening even amongst the most impracticable. But it is necessary that more signs of vitality should be exhibited by owners of property, merchants, traders, and all who are interested in the progress of manufactures amongst us;—these may show their desire for the extension of manufacturing industry in a variety of ways without interfering with their ordinary business. Landowners may, by allotting sites for mills and workpeople's houses by the sides of rivers, where there is good water-power, and by freely advertising, fix the attention of capitalists upon the capabilities of the country, as well as secure good tenants or purchasers. Strange to say, landowners, somehow, seem to suppose that persons in Manchester, Belfast, or Glasgow, may know of their properties, without their taking the least trouble to make known their existence, or the terms upon which they are disposed to let or to sell them!

On the other hand, we find the notion exists that publication may affect the price, and some, therefore, suppose secret inquiry is the way to secure suitable places at low rates. This is a total mistake. Hiding what one has to dispose of, or misrepresentation as to what another wants, is prejudicial to the interests of all parties. It may be that people on the Corrib will not let their lands cheap enough, or give sufficiently long leases. If they are so blind to their interests, then there is the Shannon, and even if the proprietors there do not put themselves in the way of business, there are water-powers and lands to be had on the Lee, the Blackwater, the Robe, the Moy; and, even though all the southern and western owners stood in the way of progress, it is only a question of canals, which might be inexpensively cut, or tramways, or railroad extension, and all the rivers in the North could be utilised to their last bucketful of water, and manufactures be carried on thus still cheaper than by steam-power.

We believe, however, that the owners of property in the more eligible districts of the country do already, to some extent, realize their opportunity, and will act worthy of it to an extent calculated to bring every place where manufactures may be prosecuted, under the notice of those persons in every part of the United Kingdom likely to avail themselves of such facilities. But valuable time is being uselessly frittered away, and the owners of property and the interests of trade are both suffering. This is not a case in which free grants can apply; yet low prices must be fixed, or long leases given, and both on moderate terms. It is not to be expected that money will be laid out in mill property if there is the least uncertainty as to the tenure.

The capitalist must feel himself independent of the landowner. The landowner ought to be sure he is dealing only with a capitalist who is able to lay out sufficient money to erect suitable works.

This is no case either for experiments on a small scale, or uncertainty in respect to what landlord and tenant are to do, or may expect to be done for them, respectively.

Perhaps the best way would be for landowners to build mills and workers' houses, and then let them. Most of those who have suitable places for such erections have also sufficient money to build them. The beneficial effects on the other parts of the property of any owner who would erect factories, would be equal to more than the highest rate of interest he might get in any other way for his money. It would also give an opening for his younger sons getting into a trade by which they might become rich, and in which they would always be as respectable as in those positions, either at home or abroad, in which those of their rank have hitherto been found. Had Connaught possessed hundreds of mechanics, and Munster been well supplied with merchants and engineers, and both had a liberal share of handicraftsmen, the flax trade (once flourishing near Galway), and the opportunity once given the people of Cork, would not have been allowed to slip.

Too great a desire to get away from home, common to all who have cleverness and know no handicraft, and the prospects of adventure which haunt the minds of all who have genius and no profession or trade, have led, more than perhaps any other cause, to the present state of Ireland in regard to manufactures. What was Ireland's loss has been England's gain, and our practical neighbours have well and faithfully improved their advantage. Negligence was the origin of all that Ireland ever lost in linen, cotton, woollen, or any other trade, but especially in the cotton trade.

Happily, however, what was lost in the South and West was not allowed to leave the country. The linen trade largely, and cotton to some extent, was laid hold of by the Northerns, and they have kept them.

But it is never too late to mend. Both cotton and linen may be brought again to all parts of Ireland. The very severity of the sufferings which have followed former negligence may, probably, hinder the recurrence of like folly again. Now is the time for action. If landowners do not build mills, let bankers advise those who have large balances to their credit to do so. Merchants, traders, and shopkeepers, who have surplus cash, might invest in this line with the greatest possible advantage to their business. If manufactures be introduced largely, the sons of shopkeepers also, who have talent, might, like those in higher ranks, become merchants and manufacturers. It is only in places where there are manufactures and trade that even the cleverest can rise. But in every such place all who are industrious do certainly rise. Character and merit, where trade and manufacture exist, must tell; but where there are no manufactures, and but little trade, character and merit are treated too often by even their possessors as worth little.

The cheapness of motive power, already noticed, in its relation to the future of the cotton trade, is more worthy general attention than any cotton substitute. Water-power, it must be recollected, is not only cheaper than steam-power, but it is inexhaustible. Even though coal be plentiful, this fact is not to be overlooked. Besides, the adoption of water-power at once raises, say £15,000, or portions or multiples of that sum to the level, in relation to nett profits on a given business, with £20,000, or portions or multiples of that sum employed in steam-power. It seems not merely enough to present to the capitalists a good case in favour of manufactures. It is also requisite for those interested—that is, for every member of society—to show, by word or action, that they desire to see capital coming into the country; and to do this, if we may be allowed to suggest the way, is, for each man to put his own money to use at home. Were the erection of cotton factories set about at once, they would not be ready a day too soon to meet the wants of that great trade which must set in after so long a time of clearing out of stocks in the hands of shippers, agents, and even shopkeepers. Factories in Ireland, worked by water-power, could not be erected as fast as they would be tenanted.

CHAPTER IV.

LINEN MANUFACTURES AND FLAX CULTURE.

THE course of emigration seems to imply that people pay more heed to far-off places, of no great value, than to places at home, which are far more valuable. The origin of this error cannot be traced to the saying that "Far-off fields look green," so much as to the unwillingness of people to scan narrowly the nature, and estimate accurately the value, of what they have long been used to look upon. This aptitude for scrutinising things, and accuracy in appraising their worth, is more the result of that general clearness of intellect which is ordinarily a concomitant of liberal education than of any faculty peculiar to a particular people. Had the Irish been as well educated in practical things as the Scotch, they would have developed, as the Scotch have done, their own resources first, and have gone to other countries afterwards. The resources of no other country are better developed than those of Scotland; but the Irish and the Scotch, who acted so differently to their respective countries, have run the same race after prosperity, and been about equally successful at a distance from home. Deficiency in general education is rapidly lessening; in like proportion is the ability of Irish people to choose the best field for enterprise manifestly increasing. Until home materials are to a fair extent exhausted, no enlightened people leave home

C

for any other country. Emigration, therefore, is either the result of ignorance as to resources in the place of one's birth, or the expression of a praiseworthy independence after these are used up.

Most Irish emigration has been the result of pure ignorance, but that of other parts of Great Britain are evidence of independence. Immigration and re-migration may be expected in this country, just in proportion as education guides people at home to the utilization of our material resources. Another and a most powerfully operating cause of emigration from Ireland has been the ideal division of the country by its natural boundaries, so that Connaught was supposed to be as far from Ulster as the United States of America or the wilds of Australia. Ulster was also as far from Munster and Connaught, for all practical purposes, as if divided by an ocean or a gulf. Formerly Ulster people went very seldom to other parts of their own country, but a great deal to other countries. America and Australia, and all other places, were the common rendezvous of people of all the Irish provinces. Many met, for the first time, in foreign lands, as Irishmen who had lived long close neighbours at home, and who never thought of their common ties of brotherhood. Happily, however, in spirit at least, these distinctions are amongst the things that have been, and it now only remains to introduce manufactures, and a platform founded upon commercial interests may become the scene of fraternal recognitions. It ought to make no difference to anyone, except that which attaches to particular trades or callings, whether an Irishman live in Ulster, Leinster, Connaught, or Munster, as in any province he is still at home. Belfast is not far from Cork; nor is Galway, Limerick, or any city or town in the island, far from any other. This is a fact rendered very palpable by railway extension. The admission of this common-sense idea would be one of the most effective stimulants to manufacturing progress.

The extension of linen manufactures, or rather their resuscitation, in Connaught and Munster, is an idea forcibly suggested by the facts now before our observation. As a means to this end, it is universally admitted that flax-growing must become more general both in and out of Ulster. But increased flax culture implies markets for its sale near the farm. The introduction of such markets in a part of this country where there are no large towns, implies the establishment of mills to spin the flax, which would create such a supply of yarns as would hasten the adoption of the power-loom on all classes of fabrics, from the finest to the coarsest. We are aware that some classes of goods have never been made by power-looms; but feel justified, on good authority, in saying it is possible to substitute the hand-loom for the power-loom in many, if not all, of the very fabrics on which some say it is impossible to adopt it. The differential profits referred to in Chapters II. and III. do not apply to linen in the same sense as to cotton. Perhaps, however, the advantage to this branch of an extension westwards would be even more than those differential profits represent. In Connaught the power-loom may be introduced without opposition to an extent impossible, in sight of the " Shanehill." The fullest possible adoption of machine labour is absolutely requisite in conveying the trade to the point indicated by the adoption of machinery in the spinning of flax, and this, too, may best be done in new mills. The mere difference between water-power or steam-power is comparatively insignificant in the flax trade, because success in that line depends so much on nearness to places where the greatest quantity of the raw material is raised. Yet the very same process of extension under which flax mills would be erected in Munster and Connaught, and parts of Leinster, implies not only more flax growing in these parts, but a general expansion of the trade everywhere.

The very first information the Irish farmer now needs in regard to flax, is to be able to detect the knavery of persons who palm off spurious seed. Such a practice, though now almost unknown in the counties of Down, Antrim, Donegal, Derry, or Armagh, still exists to some extent in Monaghan, Cavan, and to a large extent further west, to the robbery of the confiding farmer. A public officer should inspect every barrel or hogshead of seed offered for sale, and none be allowed to pass that is not sound, clean, and new. Such an officer could not only examine seed in bulk, but watch the progress of sales, and see that no tricks were played in the delivery to the farmer. In connexion with the linen trade, it seems a good opportunity for mentioning, to the credit of the Belfast shippers, that they sent their goods by way of Galway as soon as the Atlantic Company's boats were put upon the line. This fact alone shows that there is no lack of disposition on the part of the Belfast merchants to do their share in the cause of Galway's advancement to her natural position as a shipping port, from which it may be inferred that they would willingly encourage the growth of flax westwards.

Gentlemen formerly resident in Ireland, now in the United States, are getting into the habit of frequently visiting the Old Country. We believe not a few of these would stay, if they had but a fair chance of knowing what the south and west of Ireland really might be made by industry.

In connection with the cotton, flax, and woollen trades, we may mention that sales of all kinds of raw materials by auction will become necessary in our large commercial cities, such as Dublin, Cork, Belfast, Galway, &c., so soon as the trade begins to expand. Nor would the auction mode of selling then be confined to wool, cotton, and flax, but would extend to provisions and other articles as well, and to foreign grain, tea, sugar, and to every kind of imported goods.

The question of increased manufactures in Ireland ceases to be a capitalists' and manufacturers' question merely, so soon as it assumes a tangible form. Farmers begin then to feel that there is such a demand for produce as gives them the opportunity of choosing those crops which pay best. As matters are now, no such choice is in their power. Shipowners could then, also, choose the safe Irish port, or the less accessible one in England, according to the rate of freight offered. Such a choice is not now offering. On the introduction and extension of manufactures more depends than any person could now possibly predict.

CHAPTER V.

THE questions with the farmer in regard to flax culture are, not only will it give a profit, but will it give a larger profit than that which it would displace ; and, also, for what ought it to become the substitute amongst crops?

The average area under crops in Ireland is under 6,000,000 acres. Of these only about *one-fortieth* were under flax in 1862, and this was over the average of the foregoing eight years. The produce of a statute acre of flax ought to sell for from £16 to £18, even as it is generally treated at present; but, were the intelligent and highly-important remarks of a correspondent in the *Northern Whig*, of the 31st July, 1862, adopted, we believe the value of a flax crop might be raised to £20 per acre, on an average of any five years. In the letter we refer to it is said, after commending the flax bolls for cattle feeding, " By neglecting to save this part of the crop (the bolls or seedpods) there is a loss to the grower of from £5 to £8 per acre, and to the country of £500,000 to £800,000 annually, calculating the entire crop at 100,000 acres."

In the same letter the economising of the " steep water" is commended, on something like the plan mentioned by Sir Robert Kane, in his " Industrial Resources." We believe our proposal for 1,000,000 acres per annum ought to commend itself, if for no other reason than that the growth of such a quantity might become the occasion of providing tanks at or near scutch-mills, in which flax could be steeped, and the shoves put into the water as soon as made in the mill, although it seems as if this is not the most likely thing to be done with either steep water or shoves. The difficulty of finding spreading-ground in the neighbourhood of the mill, and then the cartage of the " manure" to the farm, is each a drawback to the practicability of this plan; but, taken in its worst light, its adoption would be a decided improvement on the present system of allowing both steep-water and shoves, to a large extent, to go to waste.

The rent of land is the same, whether it be under flax, potatoes, or oats, or other crops; but the payment of rent becomes a much easier matter if so much more than the rate per acre can be raised by a crop in all other respects as good for the farmer. We believe that £2 to £3 per acre more can be made of flax than of either oats or potatoes, when grown beyond a limited quantity. If farmers kept accurate accounts, and put on the debit and credit sides the cost of seed, labour, and every other incidental expense following the growth of their several crops, and then give each crop credit for its value, they would find flax to stand at least so much above all others as to entitle it to the acreage per annum we have named. It is by the profit or loss, supposed or real, much more than by the guidance of chemistry or scientific agriculture, the farmer selects one crop or another for his lands. (See Appendix A.) We know not how anyone can travel over Ireland and not be struck with the sad neglect of some fields in many even well-kept farms, several whole farms in otherwise well-cultivated districts, and in not a few cases of whole districts of country, whose so-called tillage is a mere mockery of the fertility of Irish soil, all which are puzzling to any one who listens to speeches at agricultural meetings, or judges of farming in Ireland, by the reports of cattle shows.

Something is wrong, and it requires, in our opinion, but a very superficial inquiry into the cause of this manifest negligence, to trace its origin to a mistaken notion as to which are the best-paying crops. That notion, baneful as it is, has its foundation in ignorance of the relative nett proceeds of one crop or other; and, therefore, the cure for random selections, and the perversion of lands to the growth of all but useless crops, must be by aiding the farmer to find out what crop is most profitable. When the farmer has done so, there can be no doubt that the most profitable will be chosen; and there is little fear of science refusing her aid to enable the agriculturist to get the most of that crop from the soil.

If the farmer were used to account-keeping, and separated the nett proceeds of every acre of oats, the produce of which he sold to exporters, from that he sold for home use, he would find that the value of every acre beyond 1,500,000 per annum is reduced by the price which must be taken for the quantity raised in excess of what is wanted for home use. The profit and loss idea is the sound one.

We feel perfectly justified in saying, for the reasons above named, that, apart from all considerations affecting the linen trade, it would be the interest of the farmer to grow more flax. Besides which, as a purely agricultural matter, flax culture gives so much more employment to labour that the changes suggested would create an increased demand for farm produce to feed labourers. The culture of flax would also employ all the labour now sent from Connaught, Munster, and parts of the other more prosperous provinces, to reap the English harvests. If in all other respects the ratio of increase were perfectly indifferent, its effect in increasing the sale of crops, and the providing work for the harvestman at home, ought to commend its consideration to the Irish farmer. It will also be noticed that, unless in the county of Dublin, where market-gardening and the raising of crops for a large city might interfere, we have given about the same proportion to each county according to its cultivated acreage, which is founded not on any opinion, but on reliable testimony as to soils and subsoils, as well as the possibilities of growing flax to such an extent, without exhausting the land beyond the easy possibility of restoration. In Ulster—viz., Antrim, 40,000; Armagh, 30,000; Cavan, 30,000; Donegal, 42,000; Down, 58,000; Fermanagh, 16,000; Derry, 46,000; Monaghan, 28,000; Tyrone, 48,000—338,000 in Ulster. Leinster—viz., Carlow, 14,000; Dublin, 10,000; Kildare, 23,000; Kilkenny, 30,000; King's, 23,000; Longford, 16,000; Louth, 16,000; Meath, 25,000; Queen's, 25,000; Westmeath, 20,000; Wexford, 38,000; Wicklow, 20,000—260,000 in Leinster. Munster—viz., Clare, 35,000; Cork, 80,000; Kerry, 26,000; Limerick, 31,000; Tipperary, 45,000; Waterford, 20,000—247,000 in Munster. Connaught—viz., Galway, 45,000; Leitrim, 18,000; Mayo, 45,000; Roscommon, 28,000 ; Sligo, 19,000—in Connaught, 155,000. Total, 1,000,000.

The produce of 1,000,000 statute acres may be fairly assumed to be 300,000 tons flax ready for market, which quantity is calculated to be some 255,140 tons in excess of the quantity produced in 1861, which, in a manufacturing point of view, and especially in the present crisis and future prospects of the cotton trade, could scarcely be overrated. But we need scarcely say, after the place we have given in these pages to cotton and flax respectively, that we do not believe flax can ever displace cotton either in the factory or the market. Yet the relative proportions in which these may be used ought considerably to extend in favour of flax, and *now is the time* when, for their own sakes, the Irish people ought to take action in this matter.

We believe the duty of reducing the cost of production in textile fabrics, so as to mitigate as far as possible the probable future advance in cotton, devolves on English people especially. But the case is changed where flax is the subject. In this the Irish are more concerned than the English, and ought to act accordingly. We do not say, however, that the subject is to be left in that position in which, by accident or neglect, it might fall between the " two stools." The British have as good an opening in flax manufactures as the Irish, and the Irish may improve the advantages of water-power as much as the English and Scotch ; but as it is wholesome for persons to move most in the circle they are best acquainted with, we expect more from the Ulster linen manufacturers, in respect to the extension of that trade to the south and west, than from the Manchester people. On the other hand, however, we expect more from the Manchester people, in the introduction of cotton into Connaught, than we do from the Ulster flax-spinners. If we are right in the project we announced, there is an opportunity now before the linen trade of Ulster, the like of which they never had before.

Of course, the saving in wages set forth in Chapter I. cannot apply as between Belfast and Galway, nor yet the nearness to market for foreign flax, as the nearness to market for cotton does; but if extension westwards implies what we again most confidently repeat it does, we conclude that great as the saving would be in cotton, that in linen would be no less, by the adoption of the course we proposed.

If the result, though coming in a different way, be the same, the principle we have offered for consideration as to capital for cotton mills would also do for flax factories. Flax machinery being, probably, on the average of all sorts, about three times as dear as cotton machinery, the importance of having a long time to pay off the capital, if it were to be paid off out of differential profits, would lead to a change in the table No. II., as to the rate of interest and the time it would require to pay off the cost of plant and the working capital commenced with. In every other sense this table applies to flax as much as to cotton.

There is still another aspect in which the extension of the flax trade may be named, and especially now, under the stimulating influence of the desire to meet the probable effect of higher prices for cotton in the future—namely, in relation to its invention and discovery.

CHAPTER VI.

FREIGHTS INWARDS AND OUTWARDS.

We find it necessary, at the outset, to draw a marked distinction between those harbours and ports whose natural capabilities are greatest and whose geographical position is most eligible, and those which are not quite so attractive. By this rule we must place amongst those deserving the earliest attention—first, Galway; second, Limerick; third, Cork; fourth, Londonderry; and, fifth, Wexford. According to this order, Galway harbour, whose development into a port would be the greatest of all the other boons of recent days to British and Irish commerce, is put first on the list. Under the guidance of geographical position, as well as natural capabilities, Limerick must come next, because it is the nearest of all except Galway to the great water-powers of the West, by whose application alone to manufactures Irish ports and harbours can ever become much, if anything, more valuable than at present. The further improvement of Cork, and especially the extension of her ship-building accommodation, would, as a result of the improvement of Galway and Limerick, become imperative in carrying on the trade of the country. At Londonderry there has been some-thing done already ; but, as the importance to the trade of that port has never yet been fairly realized, its claims as the outlet of a rich country, and the entrepot of foreign shipping, could not be safely postponed, those mentioned being first utilised. The chief claim of Wexford lies in its position. In a great commercial country it would be a valuable auxiliary. In Ire-land, as she is now, it could never be more than a third or fourth-rate port ; and, therefore, the interest of Wexford in Ire-land's manufacturing progress stands in an inverse ratio to its place in the classification we have adopted. The like remark applies to the other ports ; so that Londonderry, Cork, and Limerick, as ports, have more to gain by Ireland's progress in manufactures than Galway, because their situation places them under greater dependence on trade for development.

We believe that, until the ports of Galway, Limerick, Cork, Derry, and Wexford are, to a greater extent than at present, fitted for foreign shipping, those of Dublin, Belfast, Waterford, Newry, Westport, and Sligo, will never have such a trade as to accommodate which it would be necessary greatly to improve them. Londonderry has cultivated a trade with Scotland and England, because fish, cured provisions, grain, butter, eggs, and linen cloth, were her chief exports. Limerick depended also too much on provisions and grain, and, though she had manufactures, they were never expanded so as to become the means of attracting much foreign shipping to her port. Wex-ford has had no trade except that which grain, butter, and pro-visions gave her.

The port of Dublin owes much indirectly to Ulster manu-factures. Belfast owes everything that has raised her above Londonderry or Newry, Westport or Sligo, to the existence of the linen trade on a large scale, and the curing houses for pork, and the market for butter.

Ireland, to be prosperous in agriculture, must make progress in manufactures also. The least curious observer must see in the condition of Irish agriculture, as much as in the neglected state of so many of her ports and towns, the evils resulting from apathy in respect to manufactures.

The introduction of cotton manufacture into new places, and the extension of it and the linen trade westwards, are at this moment indispensable as a preliminary to the improvement of harbours, and the providing of dock accommodation for ships. Manufactures must precede shipping. Shipping must begin to come to a place before accommodation needs to be provided. The extension of accommodation and the increase of shipping always grow together. but neither can be expected except as a result of manufacturing activity.

The evil influence of the notion that manufactures are the followers, and not the forerunners, of shipping, has operated to a large extent, and most extensively to the injury of the country, and especially in the west. In view of the present condition of Irish ports, as compared with their respective natural capabilities, it is very clear that, however badly circumstanced a place may be, manufactures will lead to improvement.

It is no wonder, with false notions in the heads of people in Ireland, that time, eloquence, and money are spent in seeking Government aid to get docks made, which, in the cases of Belfast at home, and Liverpool across the Channel, and other prosperous places, have always been begun by private enterprise, which never fails to attract that attention by which any requisite public aid can be readily obtained. The support of dock trusts, harbour boards, and other organizations for the carrying on of the business of a port, must depend on the amount of dues received from shipping, and these on trade, and that on manufactures. The rate of dividend available to those who invest in this class of undertaking depends on the proportion in which the dues stand to the outlay in making docks, and in carrying on the business of the port. The number of ships coming to a place depends. in some measure, on the rate of charges. But, as the cost of making docks, &c., in any port in the west of Ireland would be small as compared with Liverpool, the fear of high rates ever interfering with shipping in Irish ports need not be entertained.

Obviously, the thing, therefore, to be done is, to centre all efforts on manufactures, until a beginning in these, on a large scale, has been fairly made; then dock accommodation will rise up naturally.

The extension of the cotton trade is the first way to provide commodities for inward freights. The expansion of the linen trade will also tend in the same direction. Every cotton mill of the size mentioned in Chapter I. will require 1,500 bales of cotton per annum. Each operative and employé must also be fed, to provide food for whom corn must be more largely imported than at present. The expansion of the linen trade, by increasing the area under flax in Ireland, will also tend to increase the imports of grain and breadstuffs, although not to the extent represented by the reduction of the acreage under grain and edible green crops. Shipping would also be greatly affected by the increased demand for foreign flax, as a result of more spinning and weaving, as well as the greater consumption of breadstuffs for operatives.

The connection of manufacturing industry with shipping is so close and intimate that, when the former is not carried on, the latter can never flourish. Sometimes geographical position may not be favourable where materials for manufacturing are abundant; such, however, is not the case in Ireland. Therefore to shipowners the neglect of the industrial resources of Ireland is no trifling loss. The non-development of these must affect their interests in that proportion in which their development would become the occasion of ships trading to a larger extent to Irish ports instead of to English, and especially so if Galway, Cork, and Limerick were the chief ports of Irish trade. The saving to ship property would be enormous by the avoidance of the Channel. The economy of time in making a voyage, on the average, including the farthest foreign ports, would amply compensate for any probable reduction in the rate of freight to an Irish port. Winter voyages to Liverpool would, in the case of Irish ports being prepared for shipping, become rare. The dangers of the Channel would be avoided by landing at any of the western, northern, or south-western ports, and the cost of transmitting merchandize by steamers to Liverpool would be insignificant, as compared with the advantage of speedy voyages, safety to ships, and lower rates of freights, as a consequence of anchoring in Irish ports.

The outward freights for vessels carrying cotton, corn, provisions, flax, to the extent required as a result of manufacturing extension, would be mostly manufactured goods, including, along with textile fabrics, pottery, glass, cutlery, and general British manufactures. These Ireland could supply, as also coal and salt for ballast. If pig iron, or even some of the other commodities for back freights, had to be brought round from Glasgow coastwise to Irish ports, the vessels carrying these could get return cargoes of iron pyrites, sulphur ore, lead ore, marble, and many other articles abounding in Ireland.

On the subject of freights, inwards and outwards, many important considerations depend. We believe correct information as to the cause of Ireland's decline in shipping is much needed as a means of setting public opinion right as to the best way of introducing improvements. The history of the past seventy years shows that rivalry, rather than co-operation, existed amongst Irish ports. The effect has been the perversion of energy and talent to the local and petty purposes of a particular city which ought to have been given to the general and momentous interest of every port, town, and city on the Irish seaboard. During the same period manufactures were allowed to decline, except in Ulster alone, under the double influence of a desire on the part of capitalists to get far too large profits, and the carelesness of operatives beyond the interests of the moment. Whilst these things were going on, England was growing strong by exercising her skill in manufacturing extension; and, until the introduction of free trade, little hope existed of Ireland recovering her lost position in manufactures. Apparently a change is, at least gradually, passing over public sentiment in respect to these matters. The manufacturers and merchants will flourish most, now that all markets are open to *good value,* who can give the best value, who can produce goods at the lowest cost for motive power, and can avail themselves of the cheapest freights, and the soonest and oftenest renew their stock of raw material, and most easily sell their manufactured goods, by reason of their proximity to markets.

CHAPTER VII.

WOOLLEN MANUFACTURES.

In the foregoing six chapters the expansion of cotton manufactures on a large scale, the increase of flax manufactures two or threefold in Ireland, and the centralization of these in the west, are treated in their relation to the bringing of capital into useful employment, as also in their certain effect upon the development of Irish shipping and the interests of commerce generally. In the chapters which are to follow, the influence of the proposed extension will be considered more particularly in reference to the providing of back freights for vessels trading with cotton and other raw material to our ports.

In the Woollen manufacture, as in that of cotton, Ireland has a trade already. She has also a history, especially in connection with her woollen trade, full of warning and instruction. No place can be visited without hearing the cry from some that Ireland has lost this trade irrecoverably. That she has not held her ground in any branch except linen, is too evident; but it is equally palpable that there are cheering indications of a recovery. Probably the chief reason why Ireland is behind England at this moment in manufactures is, that before the adoption of steam as a motive power in the production of textile fabrics, the absence in Ireland of that self-reliant determination to keep up to the requirements of the times, had operated to give the north of England an advantage which an abundant supply of cheap coal enabled her to improve. To this want of a self-reliant spirit, already so fruitful of evil, we hesitate not to attribute those sayings one hears so frequently as to England having got all textile trades, and the impossibility of competing with her in them. To ignorance may be traced the notion that cotton or woollen manufactures are indigenous to England, and that the highest aims of Irish efforts must be confined to flax-spinning. England got her trade by no patent right, but in the plain, open path of persevering industry. It was under no pro-tective measures manufacturers settled in the north of England; it was from the accident of coal, and the energy of the people. Steam-power has already so far fulfilled its mission as to raise manufactures to that position in which it is as necessary now to get a motive power cheaper than steam, as it was in the earlier part of this century to get that rapidity with the motion of machinery which steam-power gave so effectively. For all prac-tical purposes water-power is equally as good as steam-power, for every class of manufactures, not excluding woollens—and very much cheaper. The concentration necessary for the con-venience of buyers from other countries, of cotton, woollen, silk, and textile fabrics generally, into a district embracing no very large area and covered by railways, has accomplished its mission to a large degree also. Manchester, as the centre of cottons, and Leeds and Bradford as the headquarters of woollens, must still keep their ground. But buyers of goods are no more fixtures in any market than fabrics of particular kinds are indi-genous to particular places. Wherever there is cheap motive power, plenty of labour, raw material easily got at—there manu-factures may find a centre, and markets will be opened.

It is important to fix attention upon the fact that not the manufactures most likely to flourish at first sight in Britain, but those by which the most money could be made, have flourished. Within the past twenty-five years the manufacture of silk has increased most of all. This is one of the fabrics which some say is not to be moved from its older haunts. Next to this is cotton; woollen next; and linen last of all. The same prin-ciple may be seen by comparing the " exhibits" of 1851 with those of 1862, except that the comparison is not so favourable to silk. The primary question in reference to the extension of the woollen trade is, Will it pay by itself? To which we believe the answer might safely be given in the affirmative, provided it were expanded into reasonable proportions. If properly and extensively carried on, the woollen trade is as likely to grow in Ireland as any other manufacture. In the expansion already referred to, cotton and linen goods would be sold in Irish markets as in those of England, and the buyers of these would require an assortment—the absence of which would, by the way, be so much of a drawback on prices. But this drawback is not to be feared for a moment. Silk might also be pushed into something more than the very small affair it is now.. All the advantages of the "differential profits" contained in Chapter I., and of the bringing in of capital into trade, as referred to in Chapter II., apply to woollens, though not in the same degree as to cottons. The supply of home-grown wool is also in favor of Irish woollen manufacture. This is very large, and may be estimated at one-eighth the whole quantity manufactured in the United Kingdom; or equal to one-fourth the total imports. But, important as this would be, it must be put in a subordinate place (as in the case of every manufacture the raw material of which has to be imported largely) to that advantage which arises from the practicability of erecting factories near the sea-board, and the nearness of the island to the raw material producing countries. Ireland is nearer than England to Australia, the East Indies, South America; and equally as approachable, if not positively nearer, on the average, to all the other places from which British manufacturers draw their supply of wool. For the washing of wool no rivers are more suitable than those of Ireland. Even at present Ireland is a good market for woollen cloths. But, with a population busily engaged, at good wages, this market would, under the extensions already projected, double its present business. The enormous quantity of labour performed in the towns of Leeds, Huddersfield, Bradford, Halifax, Dewsbury, and other parts of Yorkshire, where the woollen trade is flourishing, contrasted with the quiet way in which work is done in Ireland, shows how much is to be done. A stranger may pass through some places in this country where there are woollen or worsted factories, and, except he be a close observer, or acquainted with the trade, he may not notice the factories at all. No more mistaken policy could be adopted than the " small scale," quiet and " easy style" of action common in Ireland, and it is especially injurious to the woollen trade. It might be difficult, without actual experiment, to convince persons used to large percentages that they would be possessed of more money at the end of seven years' trading than they are ever likely to make otherwise than by lowering their profits. It is to be feared that even if the advantages were given to the Irish woollen manufactures which the extension of cotton and linen would afford, it would still be necessary for them to reform their mode of procedure so as to come up to modern requirements. The prevalence of absurd ideas on the supposed impossibility of success so near the shadow of Yorkshire, like all other notions founded in ignorance, has a great effect in arresting enterprise. Success never rises above expectation. Expectation, if founded on facts, and striven after by industry, seldom falls short of realisation. All that is now wanted in Ireland is to get the public mind possessed with such hopes as facts justify. There is no good reason to expect the linen trade to rise to the height we have felt justified in assuming practicable, unless the way is opened by the cotton trade : nor is there any good reason to hope for great expansion in the woollen trade, if even cotton be worked and linens be neglected; nor in linens, in any case, if too much dependence is put on either foreign-grown flax to the neglect of home culture, or home-raised to the neglect of foreign.

That prosperity in the woollen trade, which rests on a preference given to Irish manufactures as such, or the fashion of wearing frieze, is both unsafe and uncertain. Unless Irish-made woollens are as well finished, as firm dyes, and as cheap as any other, no one ought to be expected to buy a yard. But as the Irish manufacturer may meet all these demands, the sooner his trade is dependent on fair, open competition, and on that alone, the better. There used to be a custom prevalent in this country of placing a "nick stick" (a jagged piece of wood) in rivers at the 1st of August, and according to the height to which the water-mark rose on the " nick stick," so would be the range of prices for corn and potatoes. It is wonderful what a power the

D

rise or fall of the river had in causing people to hold or to sell —and consequently on prices. Since railroads, more general information, and other causes operating to the spread of intelligence, have begun to counteract those superstitions; many are ashamed to acknowledge belief in them, even though they had been trained in a traditional idea of the safety of this oracle as a market guage. In the fate of the "nick stick" may be seen that of those equally unfounded ideas respecting England's power to crush, or desire to thwart Irish manufactures. The progress which has swept that superstition from society is also operating with these fallacies. The fear of English opposition keeping the labourer's hands behind his back, does more to impede sound progress than all other causes, real or imaginary. It would seem as if woollen manufacturers in Ireland erroneously supposed that the ordinary difficulties in what they have already attempted, were sufficiently onerous to deter them from undertaking the task of extension. But no one who has compared the manufacturing capabilities of the south and west of Ireland with those of the woollen manufacturing districts of England, can help coming to the conclusion that, were equal enterprise bestowed on this country, and capital invested to a like extent as in England, the woollen trade in Ireland would soon be more prosperous than in England. The present seems an opportune time for Irish enterprise in this trade, as in trade, commerce, and manufactures generally. If, for instance, the mind reverts for a moment to cotton (for it is necessary to do so for sake of emphasis) in all its importance as a leading manufacture, requiring money by hundreds of thousands, and opening new channels for the investment in other trades of sums up to millions, and offering rates of interest not hitherto known, even in Ireland, where "use of money" has been so notoriously high, the prospects opened in cotton alone are quite sufficient to strike the most careless, and should move the most indifferent to activity. Perhaps the opportunities offered to those in the linen trade are no less momentous in a monetary sense, nor the least less important than the cotton trade in their relation to the woollen trade as an auxiliary. We therefore say that not only is there no special difficulty in the way of woollen manufactures in Ireland on a large scale, but also that neither is there any difficulty in the way of those other trades, on. the extension of which the profits in the woollen trade in some measure depend.

Doubtless it will be obvious to most of our readers that we are right in keeping the cotton trade foremost. One of our chief reasons is, that cotton is the greatest in weight for inward and outward freights. We find also that wherever cotton goods are sold, there linens, woollens, silk and other goods are likely to be required to make up assorted orders. If, then, Ireland can supply cotton goods—viz., yarns, threads, and piece-goods of every kind—foreign buyers will come to her markets for all kinds of goods, including both linens, woollens, silks, &c.

There is still another reason why we believe the first efforts of capitalists ought to be directed towards cotton manufactures, because of the greater facilities these offer, as compared with any other fabrics, for manufacturing extension, and through that for the expansion of the linen and woollen trades. Even though the price of raw cotton should keep up so high as to increase the use of flax four or five fold (which we believe will be the case), unless Ireland becomes a general manufacturing country, still both the linen and the woollen trade would still have to contend with divided markets in making foreign ship-ments.

CHAPTER VIII.

SILK, LACE, AND OTHER IRISH MANUFACTURES.

AT present not more than 5,000 horse-power of water is em-ployed for all purposes in Ireland; the remaining 995,000 (for the total may be estimated at not less than 1,000,000) is going to waste. This waste represents six times the power employed on cotton, linen, woollen, and silk manufactures in the United Kingdom. In cotton, England and Scotland employ nearly 100,000 horse-power, partly steam and partly water; whilst

Ireland, in a different proportion of water and steam-power, employs about 800. In flax, about which so much is spoken of Ireland, and so little of England or Scotland, about 12,000 horse-power is used in the latter countries, whereas in Ireland only 7,500 are employed, to which, however, must be added the handloom weaving of the province of Ulster. In woollens the comparison is as 40,250 in England and Scotland is to 600 in Ireland. But in silk the comparison is still more decidedly against Ireland. Nearly three times as much power is employed on silk alone in England and Scotland (a portion of which is water-power); as there is on cotton, woollen, and flax together in Ireland.

The silk trade in Ireland is almost exclusively confined to tabinets. This manufacture would be less worthy of notice (being of small extent) were it not made much of, and patron-ized by distinguished personages. It is said that the peculiar skill of Irish poplin-weavers has kept this branch in the coun-try; but the truth is, it was too small a manufacture to be con-tended for by competition, and too precarious to be much counted upon, except by a people easily flattered that they had still one great manufacture, because Ireland bears the palm for tabinets. Like all small special trades, this one affords greater opportunities than trades of larger extent for the rapid acquisi-tion of fortunes. There can be no doubt that the sale of tabinets has been dependent upon a precarious demand, and that on fashion.

No greater mistake could be made than to suppose that a particular fabric can ever gain permanent popularity by special patronage. The only advantage of such patronage is, that it puts money into the pockets of a few. The history of those enriched by such means is, that they seldom use that money in the expansion of their trade. They generally leave a trade worse than they found it, or, at all events, do nothing for its improvement.

Too commonly, in Ireland, the only desire of traders is to gain a fortune, even though the trade should lose its power to yield another. In Great Britain the aim is mostly to keep the trade healthy, and to make as much as possible at the same time. Those whose sole aim is to gain most, gain the least at last; whilst those content to follow a wiser course, for its own sake, do really become the richest in the long run.

Enterprise and the investment of capital on sound principles would speedily and certainly bring back to Ireland hatting, shoemaking, and tailoring, on a large scale. Cabinet-making is another trade that Ireland ought to have a liberal share of. In this trade there must be the introduction of machinery and the investment of capital, as well as a sensible reduction in the prices of the various articles as between the housekeeper and the vendor. If such reforms be not introduced, the trade will fall into a mere traffic between England and Ireland; and even those who require first-class articles may yet have to go to England to get them. There is an opportunity at present in which joint stock manufacturing companies might be started, and, if well conducted, would certainly supplant what is called " slop-work," and yield a fair dividend to capitalists, as well as afford employment to Irish skilled and unskilled labourers. It is strange that so many cabinetmakers from Ireland should have to seek employment in England, whilst so much of the furniture sold in Ireland is made in that country.

There are several other considerable trades in this country in a sort of struggling state, such as lace, hosiery, ink, blacking, &c., all of which might be raised to the English standard if money were employed in them. These are all conducted on altogether too small and trifling a scale. It is satisfactory to know, however, that new trades are springing up every day, and some of these full of promise. The manufacture of sewing threads, though not very new in Ireland, is taking a fresh start. Severer trials could not have come upon it than it has suffered, and yet it is full of vigour. Water-power may, in this case, displace steam. But in connexion with thread-making, the sewing-machine, and the manufacturing of shirts on a large scale, we believe that it is wiser to abandon all trades that cannot be made profitable, and to look out for such as can, than to complain or force sales of "goods" which we are not able to make up cheaply. Gilford, in the county Down, and Chapelizod, in the county of Dublin, have both had much celebrity for thread-making. Londonderry occupies a position for its shirt factories (which now employ 1,500 to 2,000 hands) that is not equalled by any other Irish town. But when the results are compared with what has been done in England and Scotland in those branches, it will be seen that Ireland has as much to overtake in thread-making, sewing, &c., as in cotton, flax, woollen manufacture, cabinet-making, or any other line for which she has natural resources, or in which she has been excelled by others.

Capital must be laid at the base of all. To get money into these trades is the desideratum. There could be no time more favourable than the present; money is abundant and cheap. Competition in textile manufactures demands the employment of every means to keep up cheap production. Recent occurrences in the cotton trade have circulated much information in respect to that branch. The public mind is at last prepared to understand the great importance of water-power over steam-power as a means of cheapening production, and Ireland is the country for water-power.

There are other highly important reasons why the special attention of capitalists ought to be given to the textile manufactures of Ireland. The deranged state of commerce in America is sufficiently suggestive of danger to the hundreds of English capitalists who have money invested in that country. It must, however, be admitted that unless money could be transferred from America, so as to be re-invested profitably in Ireland, it would be unwise to transfer it at all. But the fact is, the manufactures we propose to extend in Ireland will compare, as to their safety and promise of dividends, with any investment in the money market. The facts already stated show that, by the exercise of the most ordinary enterprise, larger profits in the trades mentioned may be made in Ireland than anywhere else. In England there is no special advocacy required to induce persons to put money into those trades which are the staple of Great Britain and Ireland. But in the days in which England began to rise in manufactures, competition was chiefly between far-off foreigners, who were slow to introduce improvement, and the enterprising British capitalist, whose greatest aim was to deserve the highest place in his class. In regarding Irish manufactures as entitled to a like position as that occupied by Irish railways, &c., the capitalist necessarily must expect some competition with manufacturers occupying different parts of the United Kingdom. But, instead of this necessity being regarded by commercial men of large experience as any drawback upon the value of investments in Irish manufactures, it operates in quite another direction. This very competition should be looked upon as one of the most likely means by which both the old traders and the new would be stimulated to greater exertion. The question with the British capitalist who holds New World securities may be as to when he can find profitable and immediate use for his money. The larger dividend to be had in Ireland, and equal security as compared with England, and the greater safety and as good dividends as compared with America, should decide the matter.

Another question arises here which stands intimately related to the light in which a capitalist, as such, must view the matter —namely, how much of the estimated differences in favour of Ireland will continue in every conceivable position in commerce in which England and Ireland may be placed respectively to one another? The answer to this question must be, that in all probability those differences which now exist in favour of Ireland will not only last, but must certainly increase, in that proportion at least in which the increasing use of coals may affect their cost, and raise the expense of motive-power for staple manufactures. It may seem strange that we address so many statements to British and so few to Irish capitalists. But the fact is, such is the position occupied by Irish capitalists to a large extent, that for all practical reasons they must be included with English and Scotch. There is one small exception, however, in favour of the Irish in respect to flax and linen. In this line of business in Ireland it is mostly Irish capital which is used; but as the opportunities now before the linen trade indicate a scale of expansion requiring more money than is likely to be subscribed in Ireland. the linen trade, as well as the cotton, may be put in the list of good investments for

English money. A like remark applies to woollens. Our neighbours already see their way to come over to this country for purposes of trade; and, fortunately, the influence exercised by their enterprise is no less remarkable in stirring up Irish traders than in making money for themselves. We trust our people will soon see no difference between investing directly in Irish manufacture, and. indirectly in English, except in the higher dividend upon their capital to be obtained at home.

The condition of small manufactures in Ireland is decidedly better than ever before. The number of persons employed is greater than in any recent period of Irish manufacturing industry. The wages paid to operatives are good, and the chief cause of success, next to an abundant supply of raw material, has been the introduction of capital. If, in fact, the growth of the small trade of this country is now compared with that of the large trade of England, the comparison is not at all discouraging.

The tendency in England is towards overtrading, whilst the characteristic of Irish traders and manufacturers is to do too little. Bad as too little speculation is, it is a less evil than too much. It is easier to increase a trade that is too small, than to reduce one that is too large. Whatever may have been the undue regard shown to English goods, just because they were English—and whatever neglect has existed in regard to Irish, because they were Irish—the fact is, that now, at last, general information is beginning to set matters right.

CHAPTER IX.

FISHERIES—FARMS—MARINE PLANTS.

THERE is an obvious relation between these three elements of industry so abundant in Ireland. To the disregard of this relation may be attributed the carelessness of a large number of the people in farming, the neglect of fisheries, and the small esteem in which the valuable article of marine plants is held. Were farmers aware of the benefits they might derive from the riches of the sea, it would not be as now, viz. :—that the worst farming is in littoral districts. The more easily either marine plants or fish are to be had, the less skilfully are they at present used.

In coming at the great practical value in the relations of fish and seaweed to the farm, it is necessary to get a large supply of fish. Also to operate upon thousands of tons of seaweed, in order to extract from it the several valuable salts which it contains. The first question for a capitalist is—Can these be extracted profitably? The answer may safely be given, that even though the manure were worthless commercially — viz., that after it were made it would not sell at more than the cost of removing it—the profits realized on the manufacture and sale of the other products would be amply sufficient to pay a dividend of no less than 10 to 15 per cent. upon the outlay required for both processes — viz., for fishing-boats, tackle, &c., as well as for curing-houses and chemical works. But the fact really is this, that the manure so produced will sell readily at a price equal to an additional 2 to 3 per cent., at least, on the total capital, making in all 12 to 18 per cent. for money laid out, even after allowing for contingencies.

Although no estimate can be made from ascertained data of the probable quantity of fish to be taken annually, yet the best authorities agree that the limit of inedible, and those whose offal in the curing process would go for manure, can only be fixed by the numbers and efficiency of appliances for taking them. Amongst inedible fish we include chiefly dogfish, ray, skate, cat-fish, brem, and other small fish, which are either altogether inedible, or less valuable for food than for manure. The sunfish, which is no stranger in the Irish waters, and is a prize whenever caught, may be reckoned on for manure. Those edible fish which sell the best, either in pickle or otherwise (smoked or not, according to the market for which they would be prepared), are cod, haddock, ling, hake, gurnard, pollock (black and white), john dory, &c. Each of these would yield a considerable quantity of offal, and several of them a great deal of oil; so that fish, oil, and offal, would also go to make up the profit when treated, as is suggested, for manure.

Assuming the *debris* of seaweed to be an improvement to inedible and to the offal of edible fish, as manure, the next question is as to the quantity of seaweed, and whether the *debris* plus the fish would be as good manure as seaweed by itself? Judging by what is cast on shore, and by the length of shore on which it is cast, to which, if we add cut weed, the amount may be fixed at ten or twelve millions of tons annually. Many persons suppose it to be greatly less; some say the quantity is inconsiderable, and assign reasons which can only be satisfactory to those who may not have seen the coast of Ireland from Tory Island to Galway, where it is more plentiful, or have judged by the south-east and north-east coast. To those who have traversed the whole Irish coast, and multiplied what they have seen come in with each tide by twice 365, the maximum of the above estimate is not likely to seem too much. Nor will any one at all skilled in agriculture hesitate to approve of a mixture of seaweed with fish, any more than he will object to the absence from the manure of those properties in the weed which, for manurial purposes, are comparatively useless. Besides, certain portions of ammonia may still be allowed to remain, and also such portions of common salt as may be requisite, and neither the price of the manure be raised too high upon the farmer, nor the quantity of salt to be got from weed reduced below an abundance to cure all the fish required to work up the *debris* into manure. The combination of these materials is necessary in this country, dependent as it is so much upon agriculture; also the more it becomes a manufacturing country there will be the greater need for manuring the land, so that the soil may produce increased supplies to meet the growing demands of a people as they become more able to pay for food.

The west and north-west, being the places where both fish and weed are most plentiful, seem to be the most eligibly situated for commencing an undertaking on the large scale necessary to produce any commercial result. But the present slovenly mode of fishing in Ireland would not admit of a large manure manufacturing company getting a sufficient supply. Therefore, capital must be brought into the fishing business, either by private enterprise or by public companies, if but for the sake of manure. This would occasion boat-building, and that would call many other trades into operation, either at present unknown in Ireland, or but little followed. The sale of cured fish would also create new interests. Its distribution all over the country, and its export to England, would also give trade in other respects.

It matters little whether a beginning be made in the fishing department or the seaweed first, if the object in any case must be to make money—good results must follow in case anything is done to develope either our fisheries or sea-plants, or both. The capital invested for such purposes would be distributed over boat-building companies, and companies who would own and work boats; also some whose business would be to deal in manure, and others would buy cured fish ; as also those who would engage in making iodine, bromine, and other marine salts.

Hundreds of plans have been proposed for working the fisheries of Ireland, and yet they are still neglected. We have met with many well-arranged projects, perfectly sound in theory, which at first seemed to meet the case, and yet no results have followed. The best plans yet proposed are, in fact, defective, being addressed to people whose circumstances are really very different to what the projectors suppose them to be. It is assumed, for example, that, if an account be made up of boat building, cost of fishing-tackle, wages, wear and tear, &c., and that if the probable percentage on the necessary capital, after putting down a valuation for the fish, will yield a fair profit, the people of Ireland will take up the project, invest their money, and work their fisheries as the people of Scotland and England do. The mistake here has been in supposing that people can really see profits in the abstract, otherwise than in banks, or shares in railways, or public companies of a like nature. If the Irish people have not seen their way to the investment of money in fisheries, it is because they did not see how it would affect not only the money invested, but the farming interests also. Had the farmers been shown the value of fish as manure, fisheries would have been developed long ago. But this being neglected, it proved vain to try to enlist either landed proprietors, or our monied classes in projects which eventually got into bad hands, by whose unwise management they were destroyed. The question of manure supply is still open, and forms a strong reason for the development of fisheries, and the consequent saving of hundreds of thousands per annum from passing out of the country for guano, the best of which is too dear for most farmers, and the supply of which is becoming more precarious every year.

But although the chief consideration in reference to Irish fisheries is the value of their development to Irish farmers, yet as food, fish of all sorts are very scarce in every part of Ireland. There is room for a much larger traffic in the finer qualities of fish. There is also room for an enormous trade in the coarser kinds. Such, in fact, is the state of fish supply in Ireland, that in many places it is scarcely to be had at all. Small towns, from ten to twenty-five miles from fishing places, are not supplied so as to bring fish within the reach of any but the richer classes. Irish people, under the pretence of want of salt, neglect to cure their herrings, and continue to import from Scotland, Isle of Man, and Norway, to the extent of 100,000 barrels per annum. Government officials, private individuals, men of the highest intelligence, have all borne testimony to the value of Irish fisheries, which testimony is, that they are rich in every class of fish, from the smallest to the largest known in British waters. Irish fisheries are comparatively more productive than Scotch. Besides, all who have invested capital under the guidance of enlightened industry, have succeeded in every branch of fishing in Ireland.

To begin and carry on a trade for the supply of the home market would require from £250,000 to £300,000. This would be ample for boat property, tackle, boxes for transmitting the fish, and all the other stock in trade necessary to supply the people of any town with all descriptions of fish.

Now that the country is covered by railways, there is a fine opportunity for pushing fish into consumption in every town in

Ireland. If fish, as soon as caught, were sent to the interior, and that persons expected every day to receive a consignment, an auction sale might be held on arrival, say at a railway station; and the price so obtained would regulate the value at which dealers ought to receive fish from the fishermen. Those who would buy fish in that way would, for their own sakes, press sales, and send them where, in the absence of such a stimulus, they would not likely be sent. Without multiplying considerations which must present themselves to any one at the first view, it is clear that, whether to the joint stock company or the private individual, this branch of the fishing trade offers special inducements. Complaints are heard as to the English demand carrying away the salmon, lobsters, crabs, &c., and at the very same place there may not be a fishdealer on whom the lessee or proprietor of a fishery could rely for the sale of his fish, so as to be able to calculate on money to meet his engagements. The fact that the market which requires the most will get the best supply, ought to be admitted; which, if put into a practical shape, would form one of the strongest reasons possible for keeping a home supply, by giving a fair price. It is not the unwillingness of the owner to sell his fish in Ireland, but the uncertainty of the demand, that sends so much out of it. Complaints are also made of Cornish fishermen coming to Irish waters. These resolve themselves into the simple principle of competition, and if Irish people went into the fishing trade, as those from Cornwall do, the result would be that our special advantages would enable us to beat all strangers. It is not to be expected, however, that badly-appointed boats, but indifferently manned, and working by fits and starts, can compete with the best boats, the best men, and regular attention to every chance. If the paralysing expectation of Government aid had fully departed from the people of Ireland, fishing companies would soon spring up, and those now neglected resources soon be worked up to the greatest possible advantage.

A very important branch of trade might be added to all the others we have named, and that is the preparation of fish for ships' stores. This is a matter of no trifling importance, whether as regards the increase of home demand, the giving employment to the people, or the profits which it would yield on the money required to carry it on. But home markets generally depend most of all upon the circulation of money, and that in Ireland is contingent upon manufacturing extension.

CHAPTER X.

FARMS AND FARMERS.

LATELY-ISSUED agricultural returns have not been very encouraging to those interested in farming in Ireland. Something is obviously wanted to bring the aggregate produce up to the quantity to be expected from the capabilities of the soil. Even as tillage is carried on now (and it is capable of much mechanical improvement), the yield per statute acre ought to be 20 per cent. greater of wheat, 10 per cent. more of oats, and 30 per cent. of potatoes. Beans, peas, barley, turnips, &c., yield good crops, and so does flax, yet in these also there is room for improvement. It will not do merely to keep up crops to the standard of former seasons. The point to be ever aimed at is to get the most the land will yield. Nor is it enough to grow just such crops as have been grown before. The business of a farmer is, whether he holds five acres or five hundred, to grow only such crops, all things considered, as will pay him best.

The tendency to disregard the capabilities of the soil is illustrated by the increasing area under potatoes, although the yield has of late been gradually lessening, as well as in the fact that flax is not sown more generally.

Of all the wants of Ireland the want of manure is the most manifest; and of all the difficulties which impede Irish agriculture this is the greatest. This want would be largely met as a consequence of the adoption of better sanitary regulations. On the farmsteads of the small farmers there is a constant waste of manure in the very manner in which attempts are made to gather it. The manure-pit near the door of the dwelling is no less offensive to the sight than it is destructive to the quantity, and injurious to the quality, of manure. The "dunghill" is too often either sunk so as to be over-saturated with water, or is on a declivity, so that its most valuable elements are washed out. That necessary adjunct of every farm, cottage, or dwelling, should be so placed as at once to become the most convenient spot for casting out all kinds of "dirt," and also a place where any kind could be thrown without offence to eye or nostril. The observance of such very simple rules for decency would, in its money value, prove highly important, whilst, as affecting the interests of farming, it could not be overrated. At present there is wilful waste of manure to a great extent, and a corresponding deficiency in the supply of food.

It must not be supposed, however, that the mere outlay of money for guano sets this matter right. The very contrary is unmistakably the fact: the land does not get the requisite supply of manure, and never can until home supplies be economised. There is another source of manure supply vastly neglected—namely, the sweepings of markets, which are crushed into sewers and carried out of sight; whereas, these being collected, would add greatly to the common stock of fertilisers, and help to put the land into better condition. It is the business of corporate bodies to deal with this matter, although in its relation to the interests of farming, it is a question in which every member of the community ought to feel interested, simply because it affects not only the supply of food, but also the cost of living.

The universal testimony of farmers outside Ulster, as to the reason why they do not grow flax, is, that there is no demand near the farm. It might seem to some a strange thing that farmers, even in Connaught, do not see that to send their flax to Belfast, where there is always a ready market, would cost so little as to afford no excuse for not growing it.

We believe no one who has attended cattle shows would believe, except he had met with it elsewhere, how much prejudice exists amongst many farmers who occupy farms from twenty to fifty or more acres, against what they call "book-farming." On the other hand, few who know the state of public opinion on these matters, only as it is to be found amongst farmers themselves, would expect to see such brilliant successes, or hear such encouraging reports as are brought before agricultural meetings. Some effort should be made to overcome the prejudice of these objecting farmers. Let them be shown that every scientific improvement will either at once or eventually result in profit. It is also important that existing markets should be provided with every appliance for the sale and delivery of goods, with as little toil to men and waste of horse-labour as possible. We do not know whether we may not be justified in concluding that the better the "market-house," and the more convenient the pens for cattle, sheep, pigs, &c., the better the farming of the district. Probably good farming leads to good market accommodation, rather than good market accommodation to good farming. But one thing is obvious, that the markets in the counties of Antrim, Down, Armagh and Derry, are as much better, with the exception of Cork, Waterford, Limerick, Sligo, or a few other towns in the other provinces, as the tillage in these counties is, on the average, better than the tillage in those provinces. We believe the introduction of sale of grain by sample might be effected, and that it would cure many crying evils, and be found highly beneficial. We know there are many objections to it, but these could be overcome; and we believe the advantages likely to follow its adoption are so evenly balanced, as between buyer and seller, that it would be the interest of each alike to adopt it. The saving to the farmer of being able to deliver corn at convenience, rather than bring the whole bulk to market, and wait his turn for delivery, would be very considerable, and at the end of a year would show a nice little sum in money to the credit of his crops. We have heard all that can be said against sale by sample from farmers, and although there is something in everything they say, yet, as all objections may be met, we see no valid reason, after all, why such a course should not become common. It has been tried with good effect in several places, and has never failed when fairly tested.

But capital must be the beginning of all reforms in farming. Capital there is amongst our farmers; but their unwillingness to use it in their business is unaccountable. Many farmers who have money in bank, or lent at interest, go into debt for farm seeds and guano, and even do not use as much of the latter as they ought; or, what is still worse, many, for the sake of getting credit, take such seeds and buy such guanos as are the worst and the dearest. As a rule, Irish-made artificial manure is better and cheaper than inferior guano. But it is not so easily got on long credit, and therefore the unwillingness of the Irish farmer to use his money in the farm leads to two evils—namely, the paying of too much for his seeds and manure, and the discouragement of artificial manure-making in Ireland. The result is, that the ground neither yields so great a weight of produce, nor makes as good a profit for the tiller of the soil, as it ought. It is impossible to condemn too strongly the folly of those who have money, and, from whatever pretence, refuse to lay it out on their crops. It is quite true that farmers in Ireland have a good deal peculiar to the country to contend with; but on whatever principle their difficulties be estimated, they by no means prove a set-off against the special fertility of Irish soils.

The interests of farmers cannot be alluded to without opening up a great number of subjects, whose discussion has long occupied public attention. We believe that the introduction of manufactures into new districts, and their expansion where they now exist, would lead rapidly and safely to the settlement of all the land difficulties of Ireland. So soon as that general employment of labour which manufactures give would fairly set in, its effect on the value of land would be felt immediately. It would operate in two ways—firstly, in calling off very small holders to more profitable labour, and thereby changing their notions as to the value of land, on which they can now only exist at starvation point; secondly, in stimulating those who possess land to raise produce to meet the growing demand arising from greater employment, and a larger circulation of money.

Under such circumstances farms would become the size, and be let at the rents which would suit the interests of both landlord and tenant. They would then be able mutually to fix rents as merchants and shopkeepers do now the value of tea, sugar, corn, &c. Insecurity of tenure is the common excuse for bad farming in this country. This is a plausible excuse at first sight, but vanishes on close examination. The better farming of England and Scotland does more towards getting leases for the farmer from the landlord than the giving of leases does in causing good culture. The letting of lands in most respects is like the letting of houses, and, in a general way, the best tenants will be the most liberally dealt with; but the value of houses is so well known, both to owners and those who seek to occupy them, that there is no difficulty in arriving at a fair rental. The people of this country, however, do not as yet seem to know the real value of land; hence so many difficulties in adjusting the relations of landlord and tenant. The system common in most parts of England is good—namely, that of the landlord putting farm-buildings, &c., into good order and keeping them in repair, leaving the tenant his capital to work his farm. Leases are usually from ten to thirty-one years in such cases. The several systems prevalent in Scotland are also on the whole better. But neither in England nor in Scotland, any more than in Ireland, can the relations of landlord and tenant become the least more satisfactory except in being entered into on both sides as commercial transactions. The comparative insecurity of tenure in some parts of this country has been greatly exaggerated. A tenant-at-will is but a tenant-at-will even though he pay a large sum to enter his farm, and that public opinion generally admits his right to sell his improvements if he chose to leave. The Landed Estates Court still offers opportunities for a capitalist, or a company possessed of means, to purchase whole estates, and this might still be done at prices which would admit of their reselling them in lots from twenty to fifty, or one hundred acres; or to occupying tenants, at from £40 to £60 per acre; or of letting in perpetuity at either existing rents or rents fined down to 5s., 10s., or 20s. per acre. Terms of ten, twenty, or twenty-five years might be given to pay the purchase money. Such a company could make from five to eight per cent. on their capital, still possess royalties, and sell lands at from five to ten per cent., all things considered, cheaper than like lands can be got in any other country. The operations of such a company would, unquestionably, do more general good to the interests of agriculture than legislation or any other extraneous influence could possibly achieve. Such company must at first be got up by Irishmen, to show that those who know the country best have no fear of laying out their money in it, which being done, persons from all parts of the United Kingdom would invest to any extent in such an undertaking. The formation of a company for such a purpose would cost no more than £200 to £250 per £50,000 capital, and were it attempted by gentlemen in whom capitalists would put confidence, the shares of one might be all subscribed in six months from its first announcement.

CHAPTER XI.

MARINE PLANTS.

THE seaweeds most plentiful on the Irish coast are *Laminaria digitata* (or sea girdle), and *Laminaria saccharina* (or sweet tangle). There are the richest in the valuable salts, viz. :— Iodine, bromine, chloride of potassium, sulphate of potass, sulphate of soda, chloride of sodium, ammonia, &c. The kind of weed next most plentiful is *Laminaria potatorum* (or bull kelpweed); next follow *Fucus vesiculosus* (or bladder-wreck), *Fucus ceratus,* and *Fucus nodosus.* In each of these all the above-named salts are present, but the proportions of iodine especially is less than in the sea girdle or sweet tangle weeds. The commercial value of these weeds may be arrived at by a consideration of the number of manufactures their utilisation would call into operation. Iodine, for example, of which there used to be got from seaweed at the rate of from 1¾lb. to 5lbs. per ton of kelp (but from which, under a new process lately patented, there can be got 18 lbs. to 26 lbs.), would give rise to several chemical manufactures, as well as become a great auxiliary in providing employment for the people near the coast. Iodine, as it is usually extracted, is worth at present 6s. to 8s. per pound; but, if re-sublimed, would sell for double these rates respectively. Besides, this iodine can also be converted into the various iodides required for pharmaceutical, veterinary, and photographic purposes. Hitherto the only advantages the people have had from the utilisation of weed for its salts were the sums made by the persons who burned, or who boated, and the merchants who bought, the kelp for the Glasgow market. The remainder of the work was done in Glasgow. The amount of labour kelp gave, in proportion to all other chemicals operated on in these great works to which it found its way, was but small, yet it was by no means unimportant. The demand for this one product of Irish seaweed is unlimited. Its supply may be increased extensively without affecting its price in the least. But when the cost of getting it to market in the way now adopted is taken into account (and it pays well), there is obviously most palpable neglect on the part of the Irish people who have not had iodine works on their own coast long ago. In this case, as in many others, "better late than never" is the only consolation. Kelp burning has been prosecuted by the people on the Antrim, the Donegal, and Galway coasts, and on the Arran Islands in a slip-slop sort of way, these last ten or fifteen years. Some years large quantities were burnt; in other years the people were timid, and a small quantity was made. But, little or much, it was easily sold at prices varying from £2 to £8 per ton of 22 cwt. 2 qrs. All kelp had to be sent either to Glasgow by ship, or, if to the soap-boilers of Dublin and other places, at a great expense for carriage upon what these people did not want. It was like sending the merchant the wholemeal of wheat who had no sale for bran, and wanted only flour, and doing the same to the merchant who wanted bran only. To the one, flour would be comparatively valueless; and to the other, bran would be only in the way. Such may illustrate the loss by kelp-burning in respect to the supply of the various customers for the several properties of the weed. But how great must be the advantages of a process which not only saves all unnecessary cost of transit, but yields in iodine alone five times the quantity got in this country, and three or four times that got in France.

Besides, any reform in the mode of treating weed, by which the *debris* (insoluble phosphates) could be kept at home for manure, would be highly important.

But not only for the iodine, and the manufacture which would arise from its conversion at home in its various iodides, but also for the chloride of potassium, and the manufactures its treatment would bring into operation, it becomes a matter of much moment to carry on the whole process, from the gathering of the weed to the production of the most refined commodities, upon the Irish coast. The simplest principles of common sense demand that the bulky article of kelp should not be sent to Glasgow to be reduced to its elements, but that these elements, ready for their various uses, should be made fit for market. Chloride of potassium is used extensively in the manufacture of nitre (saltpetre), which is in large demand at all times. The excitement in the market for this article in times of war does not, as in the case of some commodities, lead to an extraordinary supply which in times of peace cannot be sold. But, on the contrary, the demand for saltpetre depends to a large extent upon coal-mining, blasting in quarries, and other works, which in times of peace are most energetically prosecuted. Chloride of potassium ranges from £20 to £28 per ton. Sulphate of potass, sulphate of soda, chloride of iodine, and ammonia, would each form the chief element in a separate manufacture.

Each of these could be converted into so many forms, and the existence of manufactures for each would be so powerful in sustaining or attracting other manufactures to any place, that it would be unfair to say less of the manufacture of seaweed than that it might be the means of making several towns in Ireland as famous for chemical works as Glasgow in Scotland, and New-castle in England, are now.

Viewed from whatever direction, the utilisation of seaweed is a most important matter. If agriculture is to be our chief reliance in Ireland, the economy of seaweed, especially in the way indicated by the new process referred to in a former chapter (but at home in any case), would provide the farmer with large quantities of ammonia. This necessary ingredient, which, in combination with other less valuable elements, gives such value to guano, is so largely present in seaweed, that it could be produced by tens of tons annually, and sold in the crude state at prices not more than 40 to 50 per cent. over the cost of high-priced guanos. If even the seaweed, in its wet and bulky condition, could be transferred (as it could not) to every part of the country, the effects of its ammoniacal properties would neither reach the soil to the same extent, nor be proportioned so accurately to the several kinds of soils, as if extracted and used with manures. In order, in this case, to have the aid of chemistry in agriculture, it is necessary to employ capital so that every element, in its most exact proportions to be found in sea plants, may be put into a condition in which it may be brought at once to bear upon the fecundity of the soil. The cessation, therefore, of the kelp-burning may be as beneficial to the interests of agriculture as it is requisite for the better applications of science, and the more advantageous employment of labour and capital.

By the kelp process, the state of fusion in which it is necessary to keep the weed, as may be seen, requires a heat which dissipates another of the most valuable ingredients in weed—namely, bromine. This is another reason why the mode of operation which lays hold of all the elements presenting themselves in the proportions in which they exist, ought to be performed. Besides, by the new process the commercial results are such as are likely to lead to the introduction into Ireland of those general chemical works with which the manufacture of the above-named salts must be combined.

But, in addition to all these considerations, the utilisation of seaweed presents another and a most valuable aspect—namely, the relation of such parts of its products as sulphate of soda (glauber salts) to soap-boiling. Out of this article caustic leys are mostly made; also, salt cake, or crude carbonate of soda. From chloride of sodium can be obtained (in addition to common salt) bleaching powder, or hydrochloric acid; and from the residue salt cake can also be made. But these articles, however important as the nucleus of several new manufactures, the basis of profits, and the means of calling latent capital into usefulness, are greatly more so as affording employment to labour, skilled and unskilled, and are especially to the advantage of those parts of the country at present whose people are most in need of employment. Nor are these all the elements which those invaluable plants may be made to produce, under the treatment of chemistry. There are several others, such as creosotes and oils of various kinds; but, at first, the extraction of these *would not pay.*

Had it been our duty to call the attention of capitalists to the seaweed a few months ago, we could only have mentioned the kelp process. Being confined to it, however, we could have pointed out the slovenly style with which it is carried on, as well as in the loss of freightage in carrying the mass (of which much is waste) over to Glasgow, as a good reason why a better style of kilns should be introduced, and works erected on several parts of the west and north-west coast for making iodine on reformed principles, even by kelp. Had such been all we could say in favour of this neglected resource of industry, we could have added confidently that no other class of investment available at this moment could compare with this. The profits likely to arise on the old plan improved upon might be 8 to 10 per cent. on money laid out; but by the new process three times such profits may be realized.

Besides, as kelp had to be made in the open air, and as it was not a trade sufficient to support persons who did nothing else, it could only be manufactured when the weather was fine, and farmers had no other kind of work on hands. This would have been such a drawback as would have kept the trade in bounds too small to attract much capital, or call for much enterprise. Under the new process the work can be done every day in the year, wet or dry. Persons trained to it can find constant employment, and, what is of still greater importance, it can be extended to any degree. The gain by the fixing of bromine and ammonia, as well as a better quality of the other salts, gives the excess in profits.

Ireland, in fact, possesses in seaweeds alone a field for enterprise not to be matched in any other country except Scotland, where weed is also abundant, but even there not excelled. Such are the profits to be derived from the cultivation of this industry, that even if the country were to remain as it is in all other respects, by its seaweeds alone might Ireland become famous for manufactures. These long-neglected and little-understood materials are, however, beginning to attract the notice of more than the few peasants who gather weed for manure.

Of the edible marine plants to be found on the Irish coast, carrageen, or Irish moss, is the most valuable. This is to be had in great abundance. Some of it is gathered, bleached, and sent to the English market. But the quantity gathered, as compared with what is wasted, is exceedingly insignificant. This plant contains iodine also, but only in such proportions as accounts for its peculiarly valuable properties as a jelly, or as food of any kind. Irish moss may be made most useful to invalids. The jelly made from this still despised weed possesses all the valuable properties of animal jellies, along with healing, soothing, and nutritious properties for persons under colds, and of delicate health, which are not to be found in any animal jelly. The article has never yet been fairly treated, for such are its nutritious properties, that, were they fully known, there could be no doubt its use would become general. Irish moss could be used as a substitute for rice, tapioca, sago, and even arrowroot. In this commodity alone a large traffic might be carried on.

Suppose this article were used exclusively for feeding calves, it might be made a valuable commodity of commerce. Because of its great bulk it is an article which, if not sold near the coast, and especially if it has to be sent across the Channel, costs an enormous carriage in proportion to its value. The bulk of a ton weight is no less than that of four to five tons measurement in the way it is usually packed; but, like hops and hay, hydraulic pressure would reduce its size to a reasonable bulk. Irish moss has also been used as a substitute for other more expensive glutinous substances in calico manufacture. In this respect alone its value is very great. It ought, therefore, to receive some more intelligent attention, and we trust the time may soon come when it will be required in Ireland for all purposes for which it is suited.

To any who may be acquainted with marine plants, a visit to the west would be convincing that these are valuable, and may be easily obtained in any quantity requisite for most extensive utilisation. But, if other evidence is required, the best authorities admit that Ireland has a supply, for both quantity and quality, of these resources, such as to excite astonishment at the present neglect of them.

CHAPTER XII.

TRAMWAYS.

To all further efforts after industrial progress in Ireland, tramways are indispensable. Wherever peat bog, coal, pottery clay, glass sand, marble, granite, and iron ore, are to be carried to a distance, the tramway must be employed. It is the cheapest available mode of transit. To meet the several classes of demand, the kinds of tramway to be supplied must be very various. Some cases will require the moveable sort, and others a permanent "way." In several places the "way" must be light, so that it will not sink in the ground; in other places it must be strong enough to bear heavy weights. Some of the tramways required for this country must be worked by horses; others may be wrought by stationary engines: some will have to run over roads made already; for others new roads must be constructed. In some instances artificial levels will have to be made; in many cases natural levels may be taken advantage of. A few will have to go up rising ground loaded; some will have to be constructed so that the loaded waggons running on an inclined plane will draw up "empties." Even other kinds than all these may be required. But, whatever be the class of tramway suited to any particular case, the condition of the country is such as to give work enough to make tramways pay. The cost of the several kinds necessary for such a variety of work would be widely different. Probably the range for the "way," and laying it down, would be from 7s. 6d. to 20s. per yard. But this question is by no means the most important, nor is it just now of any great moment at all, as whatever might be the cost within the range of precedent, even if 10 per cent. over and above wear and tear were demanded on the money invested, it would not make the expense of transit by tramway so high as to hinder the development of any of the several commodities named above.

Compressed peat bog would bear the expense of the most costly tramway, and still be sold at prices so much below sea coal as to gain for the peat a decided preference. But if Irish coal be worked, then the very tramways required for carrying the peat to railways, or canals would be useful in carrying coal also.

Suppose the localities of flat peat bog be taken as the guide for laying down tramways (to carry peat to market), not only coal, but minerals of various sorts, could also be had in such places as would bring them within the range of the tramway, either at easy cost for sidings, or other mode of carriage. Besides, if the transit of the produce of bogs on hills had to be provided for, there also the same tramways could be used for marble, &c. But in present circumstances we believe the safest guide for tramways is the bog. Condensed peat is now in demand, and as there is no lack of supply of bog, if it were only worked, the traffic attaching to its utilisation would give employment to hundreds of miles of tramways. The lessees of coal mines in England have long since availed themselves of tramways of a variety of kinds. It seems rather strange they have not been more generally adopted in Ireland, seeing they have already been successfully used, even in conveying bog in the wet state to a suitable place to dry it. The advantage likely to accrue from tramways at the bog is not yet known. The cost of such tramways as would suit either the coal-pit or the bog consists of the value of the land over which they run, the wages paid for making them, the price of timber for sleepers, and the iron work. In all of these Ireland has the advantage over Great Britain, except in iron; and the disadvantage in that is so little that the general result is decidedly in favour of Ireland as to the probable cost of constructing such tramways.

In England, where the distance is long enough to require a railway, there a railway (made and carried on by the coal proprietor or lessee of the coal-pit) is made to supersede the tramway; and such would, in process of time, be the case in Ireland, were tramways introduced in the working of bogs for fuel. The employment of the timber necessary to construct tramways would be a very important thing for the growers of larch and other firs. The chief of the woods in this country are in the west. Already many attempts have been made to carry on a trade in green timber between the growers in Ireland and the railway contractors and coal-pit owners in England. These attempts have failed, because the owners of Irish timber would not generally cut it into such lengths, &c., as the buyers required; and the freight of whole trees across the Channel increased the cost so much, that Irish timber has never stood fairly in the English market. This has been a greater loss to Irish growers than to the English railway contractor or coal proprietor, who wished to push a trade in this article in Ireland. But the laying down of tramways would at once create a home demand for trees, and provide a means of shipping timber to England.

Still more, the making of tramways for the purposes of peat bog development would work up other kinds of firs than larch, and also several kinds of timber besides firs. The woods of the west, and even those in the centre of Ireland, are sadly neglected, and far less made of them for their owners, and consequently far less employment afforded, than they might yield. A home demand is, in our opinion, the first step towards so increasing the value of timber as to encourage its growth in this country as in Scotland. The immediate relation in which the construction of tramways stands to timber, and the relation in which timber stands to the employment of waste land suitable for no other purposes than growing timber (and exactly suited to that), place the matter strongly before the landowners of Ireland, and especially those in the west. Calculations based on the safest data prove that even were there no peat bog, the development of coal, marble, granite, &c., would require and might be made to pay for the construction of tramways.

When it is obvious at first sight that a " way" laid from any of the localities of coal, marble, granite, glass sand, or building stones could be used for other purposes, the tramways might become as valuable to capitalists as a profitable investment, as for other reasons to landowners.

Tramways, in the present condition of Ireland, are just as indispensable as it would be to provide house accommodation for operatives, were factories started for the sake of the water-powers, as at Belleek (for example), on the Shannon.

We do not see how such material as the glass sands of Donegal can be either shipped or manufactured unless tramways be brought into operation. To ship the valuable quartz of muckish would be impossible in the present state of land-carriage. To ship this valuable article at all, except in a manufactured state, would certainly be the least profitable thing that could be done with it. But whether to sell the sand as it is, or the glass it would produce, the tramway is imperatively required.

How, also, except by a tramway, could the splendid stones at Dungiven (Derry) be got to the railway or the Foyle, for transmission to our towns or cities, for building purposes? Nor can Irish marbles, especially those of Connemara, ever be brought out properly till there are tramways in those places where railways could not be made, and where the carting is too slow and too dear. The principle of the tramway is adopted in many parts of Great Britain, and a few of our own large cities, even in the inside of warehouses. The pushing of weighty goods from one place to another in trucks can be done by the touch of a child. But, strange to say, up to this date, many important parts of the business of Ireland are either done badly or neglected altogether, for want of these now common and necessary adjuncts of trade and commerce.

There is another class of tramways, however, equally important—namely, tramways for roads, in substitution for the horse, coach, van, or cart. In the west especially these are much needed. Several parts of the north require them, too. They are much wanted for Kerry. This would, for example, be a cheap, safe, and satisfactory mode of testing the wants of a backward district before going to the expense of a railway. The facilities offered by this conveyance would, in fact, open up many a place now almost closed to trade. A visit to the coast from Galway to Clifden, and the several districts outlying Westport, Ballina, Sligo, and Roscommon, would show the want of the tramway. Wherever there are fisheries, especially, the tramways ought to be introduced.

The people in Connemara make the material for most of their clothes out of wool. They also make some coarse linens, but if they had a market for their wool and flax they would, as others do, use factory-made goods, which are cheaper. The transit of the wool of Connemara would itself (whether it were carried to the seaboard for shipment, or the river-side factory) afford considerable carriage for tramways. There are as many sheep raised in the district as might produce wool equal to 6,000,000 lbs. per annum. The primary importance of this fact is, doubtless, in relation to home woollen manufactures. But, in a secondary view, it is no less important to tramways, as, whatever be the use to which the wool is put, it must be transferred from the farm, and the cheaper the rate, and the quicker the speed, the better.

There is another district in Donegal outlying every town beyond Lifford northwards, and including Donegal westwards, in which, so far as levels admit, there ought to be tramways for the very same reasons, except that in the supply of wool Connemara is peculiar. In many places in Munster, not far remote from towns of considerable size, there is a torpor which is never likely to be eradicated until a movement be made in the direction of road tramways as stage conveyances. Much may be done in this work by private enterprise. We believe, however, it would be done best of all by public companies, but not by Government grants in any case.

CHAPTER XIII.

PEET BOGS.

Of all the sources of industry in Ireland none deserves a higher place than peat. In the immense quantity of this most valuable article in the bogs of Ireland there is the means of industrial power, and consequently of wealth, beyond all calculation. The bogs most valuable in an industrial sense lie south of Dublin and west of the Shannon. These would yield fuel for many years to come, and this fuel for domestic use would be both cheaper and more easily obtained than sea coal.

Many projects have been started for getting its valuable products from peat. The experiments designed to utilise peat as fuel have succeeded best, and the present circumstances of the country afford a suitable opportunity for the further adoption of peat fuel, if not its substitution for coal for house purposes. Coal usually ranges at from 15s. to 20s. per ton, in towns near the bogs of the west; in which towns peat fuel for domestic use might be sold at 10s. to 12s. per ton. The outlay of a few thousand pounds in making "ways" to the bogs, and erecting machinery for compressing the peat, may bring this article into every town in Ireland. The utilisation of Irish bog, to the production of an abundance of cheap fuel, must be conducted on a large scale, and in something like the way English and Scotch people go into coal mining. The preparing of peat for fuel might, in fact, be made a most extensive source of industry. Not only might the towns of Roscommon, Galway, Mayo, Limerick, &c., be supplied with cheap fuel from the western bogs, but the price at which peat fuel might be sold would bring it largely to Dublin, and just as coals are sent to various places now, peat fuel might be sent to displace their use. The several smaller bog districts would also come in for a share in trade, just as the several smaller coal districts in England get a part of the general business. In places approachable from each bog there would be a brisk competition. Besides, the variety in quality of the several bogs would produce a list of descriptive names to guide purchasers, just as those attached to coal. Manufacturing in this article might be carried on every day in the year. Even if stocks of peat fuel were not gathered in several towns against severe weather, it would be easy for merchants to get supplies at a day's notice, so that *famine prices for fuel* would not be likely to occur often, if at all, even in the severest season.

To put a supply within reach of the poor, who often suffer from the want of it, seems so much a duty that but few approach the subject of Irish bogs without feeling in some degree in danger of meeting with an appeal in behalf of the poor. Therefore, many who willingly invest money in coal mines in England, or engage in the coal trade to make money at home, never think seriously about the possibility of getting a larger interest for their capital by putting it into " bogs," and making more by dealing in the home-made peat fuel than they can in sea coal. If the money belonging to Irish capitalists, directly and indirectly employed in English and Scotch coal mines, were transferred to Irish peat bog manufacture, it would be nearly, if not fully sufficient, for their development, and would pay the capitalist at least 2 or 3 per cent. per annum more. Besides, so long as the business carried on in fuel is confined to trading in, or even importing coal, it can never afford either such scope for commercial enterprise, or give such reward for labour, as if like enterprise and industry were employed on a home-made article. English coal would really have no chance of beating manufactured Irish peat fuel for Irish consumption, provided both could be had in such quantities as would give consumers a chance at all seasons of the year.

An export trade to no inconsiderable extent might also be carried on in peat fuel to America, and even in England. But, as the supply of our home demand is so important, it would be for a long time the most profitable. Farmers would gain most by an increased supply of cheap fuel. Many articles of cattle and pig feeding require to be cooked, and the cost of fuel enters largely into the question of using certain kinds of feeding.

Indian corn, where fuel is cheap (in Lancashire and Yorkshire, for example), is largely employed for cattle, pigs, and even horses, always steamed or boiled. Egyptian beans, and several other kinds of hard grain of great value in feeding, may be greatly improved by steaming or boiling. These are most in use where fuel is the cheapest, and churning, threshing, and several other works about a farm, are done by steam-power (or might be, if fuel were cheap and plentiful); and, therefore, to farmers, whether large or small, to dairy-keepers, and all who feed cattle, pigs, or beasts of any kind, cheap food is a boon. But not only for these uses, but also and most particularly for manufacturing purposes, peat fuel is decidedly better than coal. Several manufactories requiring cheap fuel might be started in Ireland for the sake of the peat.

The diffusive character of heat by peat makes it highly valuable for some purposes, and the greater absence of corrosion on the outside of boilers heated by peat, as compared with those heated by coal, makes it more valuable than coal in this respect also.

Although it is well known that the same money's worth of peat will make more steam than that of coal, and especially if mixed with anthracite coal (which is abundant in Ireland), still it is strange that the notion exists amongst those who know little of Irish resources, that Ireland has been outrun in the race of manufactures for want of fuel! The fact is, had the peat bogs of this country, situated as they are so favourable for the western export trade, been economised, Ireland might have entered into successful competition with England in the cotton and woollen trades, even by steam-power. In such a case Irish fuel would have been as plentiful as English, and mills worked near the bogs of Connaught might have had their engines fired as cheaply as those in the Manchester coal-field. But as water-power (as already shown) is so much cheaper than steam-power, however cheaply generated, there is in the apathy of our people in regard to the utilisation of peat for steam purposes the less to regret, as it may yet be used for other purposes. Steam-boats on our lakes and rivers are becoming more numerous. These are generally worked by peat, even in its bulky state; but peat fuel also suits railways. Testimony in regard to this, as founded on experiments recently made, is very decidedly in favour of peat, as compared with even the best qualities of coal. After a trial on the Belfast and Northern Counties Railway, it has been certified by Messrs. Alexander Yorston, locomotive engineer, W. H. Stephenson, engineer, R. D. Domville, resident engineer, all of Belfast, " That, having carefully noted all facts, we have no hesitation in saying that we consider the condensed peat in every way well adapted as a fuel for locomotive purposes." At this trial 14 cwt. 1 qr. 14 lbs. of peat fuel was used to drive a train (including engine and tender) of 70 tons, 74 miles. The speed attainable at any time the pressure was allowed to rise to 120 lbs. was 40 miles per hour. Experiments made on marine engines are equally satisfactory; so are those made at corn-mills or factories of any kind. The utility, therefore, of peat in a condensed form for steam purposes, is admittedly even greater than that of the best steam coal. In fusing iron, also, peat is superior to coal. This has again and again been proved by experiments; but, as the Irish peat bog question has been allowed to rest, and, as it is most desirable that all projects for its future utilisation should stand on authority which cannot be gainsaid, we give the following quotation from a report on five samples of peat such as might be had in fifty places in Ireland, and to an extent to justify the adoption of the most improved machinery, and the erection of the most permanent works, for its condensation.

Boiling, evaporating, and fusing Properties of five samples of Peat Fuel as compared with Coal, the same Quantity of each in Weight being used.

Fuel	Time in which the same body of water was brought to boiling point.	Time in which the same body of water was evaporated.	Time in which complete fusion was effected.
Coal, good furnace	6 minutes.	14 minutes.	31 minutes.
Peat, No. 1 ...	1½ ,,	6 ,,	14 ,,
Peat, No. 2 ...	1 ,,	7 ,,	17 ,,
Peat, No. 3 ...	1 ,,	7 ,,	16 ,,
Peat, No. 4 ...	1 ,,	6 ,,	17½ ,,
Peat, No. 5 ...	1 ,,	5 ,,	12 ,,

All the samples of condensed peat tested in the above experiment were produced from the same bog, and were of fair average quality. They were treated differently, having been subjected to different degrees of heat in the process of drying; consequently the difference in their results is due to the mode of treatment, and not to any difference in the quality of peat. It will be observed that each of the five samples of peat raised the heat of water to boiling point in about one-sixth of the time required when using coal, and that the same quantity of water which was evaporated in fourteen minutes by coal, was evaporated by the same weight of peat in five to seven minutes. It will also be noticed that metal which required thirty-one minutes to be fused by coal, was fused by one sample of peat of the same weight as the coal in eleven minutes. The duration of the other samples of peat was one-third less than that of coal. The relative value of the fuels, taking into account the difference both in their duration and in the duty done by them (but irrespective of the difference in the time required for evaporation or fusion to commence, which is greatly in favour of the peat), are represented by the following figures, viz.:—

Coal	...	...	...	1·00	Peat, No. 3	...	...	1.33
Peat, No. 1	...	...	1·55	Peat, No. 4	...	...	1.55
Peat, No. 2	...	...	1·33	Peat, No. 5	...	...	2.80

The No. 5 sample of peat was in the proper condition for generating steam, or for smelting iron or other metals. The sample No. 1 was best suited for domestic fuel; Nos. 2, 3, and 4, were in the best state for the production of gaslight.

The above experiments were made in June last by Messrs. Jackson and Johnson, operative chemists, 89, Bishopsgate-street, London. Valuable as this testimony is, it is only corroborate of the results obtained by Irish chemists.

There is also another fact in connection with peat which has been brought under our notice by a practical chemist, well acquainted with the Connemara peat, that there is in Connemara peat capable of being dried, for all practical purposes, to the same hardness as average condensed peat.

For gas, also, peat has been found most valuable. Mullingar gives a clear proof of this; so do several other places where peat has been adopted instead of coal. Peat has been ranked, on the best authority, in the following position for illuminating power, in reference to coal—viz., 1st, boghead coal, 4·35; 2nd, Lesmahago coal, 2·67; 3rd, peat, 1·90; 4th, Cannel coal, 1·75; 5th, Newcastle coal, 1·00. Boghead coal yields (per ton) 15,000 cubic feet of gas, of $37\frac{3}{4}$ candle light; Lesmahago coal, 13,350 feet, 26 candle light; Wigan coal, 11,420 feet, 20 candle light; Newcastle, 10,375 feet, $12\frac{1}{4}$ candle light; peat, 11,000 feet, $22\frac{1}{2}$ candle light.

The above figures are the result of experiments made very lately. The charcoal of peat is also much more valuable than the coke of coal. Peat coke is more free from sulphur than that of coal. In the article of gas, the adoption of peat, for the sake of better gas, would be a very important matter in Ireland. As a whole, the people of this country have to burn a gas much inferior to that used in England—the gas used in Scotland being better than either. The high price of coal in Ireland might be some excuse, if there were no bogs; but so long as there is inferior gas in any town in Ireland, and the bogs are neglected, it must strike the least observant that there is inexcusable blame somewhere. Most of these facts have often been mentioned before. But at least one thing new, however, may be said on this subject—namely, that the notion, once so commonly entertained, that bogs ought to be used for other purposes than fuel, has now almost died out.

It is now required that capitalists should turn their attention to machines for compressing or condensing the peat. Works should be erected for that purpose, and tramways laid down to convey the fuel to market when made. This is not a thing to be done by halves. Peat must be gone into extensively. None but large capitalists or joint-stock companies are fit for such a work.

Peat bog is, probably, the only material of all the neglected resources of this country which can be worked as profitably (though it may not be so extensively), whether there are manufactures in the country or not. Fires must be had for every dwelling, which must be the chief source of demand at all times. Next to this, gas must also be had.

In the carrying out of any peat-fuel project, a market for its sale ought to be in every town, especially in Dublin. There it ought to be disposed of by sample on representation, as coals are now in London in the Coal Exchange. Manufacturers would find it necessary to sell in large lots to merchants, these to dealers, and dealers to retail vendors.

Some people talk about engineering difficulties in respect to Irish bogs. These resolve themselves into a question of capital. Others say that it would be better to drain the bogs and reclaim them, and let the peat alone. These cannot surely have made themselves acquainted either with the special character of Irish bogs or the value of peat, much less with the peculiar nature of those deposits so like Irish bog, which are now being drained at places in England (Chat Moss, for instance), and which deposits, not being worth the cost of economising for fuel, are utterly valueless, especially in a coal district.

CHAPTER XIV.

BACK FREIGHTS.

The importance attaching to the four materials of glass-sands, pottery-clay, coal, and iron, consists chiefly in their providing back freights. If there were no sands in Ireland out of which glass could be made, or no clay for pottery wares, or coal, or iron, the special suitableness of her soils for flax culture, and the peculiar situation of her ports for commerce, would mark out this country as fit to become a rival in all modern commerce for England. But possessing, in addition to all other advantages, these materials in abundance, there are the strongest inducements to manufacturers to invest their capital here. If cottons continue the chief article of textile manufacture, then the products of sand-clay, &c., will, as heretofore, form a heavy part of the cargo outwards of cotton ships. But if flax becomes the leading fibre, then the products of glass-works will be also wanted to make up cargoes for corn ships. It must not be lost sight of that, if flax is to be cultivated in Ireland to any large extent, its cultivation must displace cereals. In that case more corn must be imported, and back freights found for ships so employed. To grow a quantity of flax equal to the substitution of cotton would require ten millions of acres, or twice the whole area under all kinds of crops in Ireland. Even if the price of flax were so high, and the demand so large, as to cause the adoption of every possible method for its growth in the least likely places both in and out of Ireland, the special character of the climate of Ireland would still mark her out as the natural flax field of the United Kingdom.

It is not the least likely that flax (or anything else) will ever fully prove a substitute for cotton. It is even less likely that Ireland will be turned into a flax field, as some would seem to expect. We believe, therefore, that the value of sand, clay, iron, and coal can be but very inadequately estimated until their value in relation to back freights be found out. Ireland possesses good quartz sand and pottery clays, iron ores and coalbeds, in such quantities, and at such available places, as would enable her to compete with England in all articles to be shipped by the vessels which usually trade in the markets for cotton, wool, silk, corn, foreign and colonial produce " inwards ;" and in yarns, piece-goods, and all kinds of British manufactures " outwards."

The chief localities of quartz sands suitable for glass of any kind are in Ulster and in the county of Donegal. There is quartz sand in many parts of Donegal, but Muckish, being the richest and most central, is naturally pointed out as the seat of manufacture for that part of Ireland. There is also quartz at the Twelvepins (Connemara), also the north of the Arran Islands. Manufactures might possibly be established in Mayo or Galway, or even in Connemara, as that district, including Tuam, has many large deposits of this valuable article. Quartz is also to be found in the county Wexford, in the Sugar-loaf Mountains, Wicklow, and at Howth ; but, on the whole, the north-west contains the most. It is there most favourably situated also ; and as it is to be found in various parts from Innishowen Head to Donegal Bay, so that for convenience' sake, glass-making ought to be in or near that district. Flints also abound in Antrim, which is another reason in favour of the north or north-west for glass-making. This locality would suit any port on the coast from Cork to Londonderry, as also Dublin and Belfast, or any other place to which there is railway or canal communication. But the question is not (in relation to back freights) how far the comparatively light goods have to be carried for shipment; but it is, how near can the works be got to the place where the quartz is most abundant, and yet be convenient to a port where shipping comes in extensively?

Whether the shipping of a country producing, as Ireland ought to produce, enormous quantities of manufactured goods, is more likely to be benefited by the nearness to ports of materials fit to produce heavy " freights," or the manufacture of such materials is likely to be benefited the nearer it be to good ports, are questions too difficult to answer except by actual experiment.

Nor are the clay districts of Ireland any less favourably situated for export manufactures. Already experiments have been made in pottery works. At Castle Caldwell, for example, some of these are going on prosperously. Some have said that pottery manufactures cannot be carried on in Ireland. This notion is but the application of the popular fallacy that accidental concentration in a place necessarily implies undisturbed continuance; whereas the most cursory observer must see the tendency of manufactures is most to those places where the raw material is to be had cheapest; in so far it is compatible with sites for works and markets for goods.

The question to be answered in respect to the export trade of pottery goods is, where are the clays to be had, and can they be worked cheaply? as, also, are there permanent " ways" to get them to the seaboard? Pottery clay is found in every part of Ireland. In some places, as on the Antrim coast, near Larne, where flints are also most plentiful, there is a very large supply of clay. In the county of Mayo, also, and in parts of Galway, there are numerous patches of various kinds of clay. But there is no part of the country in which large manufactories may be so cheaply conducted, nor from which " goods" may be so readily got to the seaboard, as from Castle Caldwell. At this place works are at present in operation, to which there is also railway communication. Some place near this part may evidently be made the great centre of Irish potteries, because it is accessible to any part of Ireland.

Iron ore is also to be had at many places within easy distance of several ports in the island. It is not, however, a matter of much moment whether there are large or small supplies of ore. The grand question is, can iron and steel works be carried on so as to supply back freights ?

The abundance and cheapness of fuel are our greatest auxiliary. Above all other fallacious notions, that of Ireland being kept back for want of fuel, has been the most fatal. After what we have stated as to peat bog, it must be admitted that Ireland is well off for fuel. We know that most people consider Ireland's supposed want of coal a fatal drawback upon her prospects. If, however, the case be fully considered, it will be found that in the branches of glass, pottery, and iron manufactures, Ireland may take a higher place side by side with Great Britain, by using her peat, than if she had as much coal as England.

The pottery district of England was not chosen because of its nearness to the clay by which the works were to be carried on, but because of its proximity to Manchester, Leeds, and Liverpool. Quartz is brought from Cornwall, clay from Dorset and Devon, flints from Kent and Antrim, and yet the great centre of manufacture is in Staffordshire, to which the approach, though now easy, was never natural, and has actually cost enormous sums of money to make. Suppose, on the contrary, some place in Donegal (Letterkenny, for instance) be chosen as the centre of glassworks. To this may be brought by water carriage any of the quartz, flints, or other materials necessary, besides what might be brought by a tramway from the neighbouring district, where quartz is so plentiful. In or near this town, and in many parts of the district surrounding, the facilities offered by Lough Swilly, or the trifling cost of tramways, or even railways, would be, as compared with the outlay made for the Staffordshire works, very trifling.

If the centre of potteries be in the district near Lough Erne, or command the seaboard at Donegal Bay, there also would the situation be better for the raw material than in any part of the pottery district of Staffordshire.

In respect, therefore, to supply of raw material, fuel, or the situation of places suitable for manufactures with reference to the seaboard, Ireland can compete with England, and, therefore, the question of back freights is in a position to be settled.

But we find the prevalent spirit is to search for excuses for the country being so backward, rather than to seek for reasons why the future may be more prosperous. To this error may be traced the notion that Irish manufacturers must be burdened by higher freights than those to which English are subject. Such chimeras floating in the mind keep persons from looking at facts. It is, therefore, indispensably necessary for all who have an interest in Ireland to consider the question—Can she supply back freights? We are convinced she can, and point to her several resources of industry as proof. Ireland ought not to be an importer of either coal or salt, but an exporter of both. If glass, pottery, &c., be fairly worked out, no one could doubt the possibility of Irish manufactured goods becoming as well known in foreign markets as English or Scotch, nor questioned the probability that it would be no drawback to their value that the ships which carried them should "hail" from Irish ports. The carrying out of the glass, earthenware, iron, and steel trade to the extent necessary to meet the desideratum of back freights would give profitable employment to thousands of persons, and tens of thousands of pounds sterling—now idle.

CHAPTER XV.

GLASS, EARTHENWARE, AND IRON.

DOUBTLESS the test to which we have already submitted the value of the quartz sands, pottery clays, and iron ores of Ireland, in confining their utilisation to the supply of back freights, will appear to the most scrupulous sufficiently severe. We feel, however, perfectly justified in arriving at the conclusion that to work these (especially as there is abundance of fuel in the country) for export alone would pay handsomely.

But the necessity for back freights, and even the practicability of carrying on any export trade at all, depend on the further extension of textile manufactures in this country, and on this almost exclusively.

Probably there are some who still suppose that Ireland will not take advantage of her present opportunity. Such a supposition is contrary to indications already observed ; for, although the growth of commercial progress is slow, it is already healthy and full of hope. Supposing manufactures to be extended, there would arise a home demand also for iron goods to a large extent. There are also, probably, some manufactories in which steam-power might be used more advantageously, all things considered, than water, provided peat fuel were manufactured extensively, steam boilers and engines for such factories might, therefore, be in greater demand. But such cases as would justify the outlay for boiler and engine to do work which might be done by a water-wheel, are too rare to be largely calculated upon. In any case, the quantity of machinery, &c., required for each factory would create a demand sufficiently large to justify the erection of more machine works of the various kinds necessary to make both water-wheels, machinery, and inside machines.

We do not for one moment assume that persons about to erect factories ought to buy only Irish-made machinery, or employ none but Irish engineers; but we do know, as a fact, that there are both engineers and machinists in this country who could do such work as well and as cheaply as those of any other country; and, as the cheapest and best workman ought to get the job, be he Irish, Scotch, or English, we are certain machinists and engineers in Ireland could make larger profits in supplying Irish factories with every appliance for any kind of manufacture than those of Great Britain could do. There can be no doubt, moreover, so soon as more factories are begun, many persons will come into this country out of England and Scotland to carry on the iron trade; and the manufacture of peat fuel itself, were there no other reason, would justify the conclusion that Ireland must yet be a great iron country.

But, supposing there be no factories erected, and no textile fabrics to be shipped, or no corn ships to be freighted outwards, as a consequence of increased flax-growing, how, then, would the glass, earthenware, and iron manufactures of Ireland stand? We venture to say that any one who may go into the subject carefully will be convinced that, if Ireland could by any possibility remain as she now is in respect to manufactures, still these works might be carried on profitably. The first evidence in favour of this assumption is the abundance of peat fuel, which may be used for glass works, potteries, and iron smelting, &c., and for that alone, except as house fuel, which in the progress of the reformed modes of cooking, and other changes, is likely to be universally used. Although the consumption of fuel would be very large in iron and pottery works, the supply in Ireland is fully adequate. But it would certainly be less expensive for glass and pottery manufacturers in Ireland to import coal from Scotland and England than it is for the glass and pottery manufacturers in England to carry their materials from various parts of England and Ireland; nor would it be any more expensive to import clays or other materials from Germany and elsewhere to Ireland than to England. The supply of these raw materials, which neither country itself possesses, being as easily got to Ireland as to England, and labour, skilled and unskilled, being alike available in either place. Ireland may surely place her special advantages in supplies of clay, &c., over against the concentration already attained in England, and enter into competition for a share of the home trade. That England has made great progress, and now has a well-earned fame for every class of article of glass, porcelain, china, delft, draining-pipes, encaustic tiles, &c., is a palpable fact; but that Ireland has materials, by the manufacture of which she might excel England in each of these, is equally apparent.

The next thing to be considered is the extent to which home demand may be relied upon. It is only the first few years that may be feared. If these could be tided over safely, the remainder of the work would be easy. It must not be forgotten that, in relying entirely on home (that is, Irish) trade, an extremely unfavourable case is assumed; and yet, as it is a critical subject, and the issues are so serious, it is safest to take the very worst possible view of the matter in all its bearings. At present Ireland is one of England's best customers for glass and earthenware goods. If Ireland had become a manufacturing country in these lines, some trade might still find its way from this country to that; but it is just as likely there would be as much finding its way from that country to this. People now-a-days buy in the cheapest market. Ireland's facilities fairly worked out would enable her manufacturers to offer the best value to be had, and, therefore, obtain from all unprejudiced buyers a preference.

Nor would Ireland be any less able to compete with England in the kind of iron "goods" used in housekeeping than in those required at factories, and for the same reason—namely, the advantage of cheap peat fuel, which, so long as it is not counterbalanced by any drawback of equal amount (as it certainly is not in the comparative price at which iron could be had in England or Ireland), is quite sufficient to decide the case in favour of Ireland. No other country can supply peat fuel so cheaply as Ireland. No other kind of fuel is so good for iron works and metallurgical operations generally. The matter resolves itself into a question of special advantages on each side respectively; and as the peat fuel, as already assumed, is worth more than "centralization," Ireland has, for this reason, if there were no other, the means of successful competition. The first source of demand is in the ordinary wants of housekeepers. This draws off every day large quantities of delft, china, crockery, pots, pans, knives, forks, &c., even in the worst of times. But, supposing the population to improve in taste, articles now considered luxuries would become necessaries, and the demand, at present increasing steadily, would grow rapidly. The fact is, the supply in Ireland of all sorts, from the most indispensable to the most ornamental, is too small. Some who have observed the small supply sent to Ireland have supposed the Irish people careless of the convenience their possession gives. But such a notion is disproved by the fact that the aggregate cost to housekeepers is nearly the same for the little they have in Ireland as it is for the greater quantity used in England and Scotland. This fact, otherwise unworthy of notice, becomes to the prospects of "home demand" a valuable consideration. To raise up the supply in every house in Ireland to the standard of necessity would require as many pots, pans, kettles, knives and forks, fenders and fire-irons, &c., &c., as also glass and earthenware of every description, as would employ a large number of works for many years. After all that may be said about the reasons why such and such works ought to be engaged in, profit and loss is the primary concern, and that depends on demand, and in this case, for argument's sake, on home demand. Besides the necessary household supply of those articles (which ought to be made cheaper in this country than where they are made), there is also a demand for trade purposes.

In glass there are several kinds of bottles used by different classes of traders, and these made at home would go a great way in giving employment. All shops, of every kind, have less or more glass bottles, earthenware vessels, and iron utensils. Plate glass, now in very general use, might also be manufactured in Ireland.

It seems strange that every specimen of the tools, utensils, &c., used by our tradesmen and shopkeepers, with the exception of a few, are English or Scotch make; and those very men that use these are laying up money, building house property, or investing in English schemes of every kind, rather than laying out their savings in the manufacture of the things they must buy daily, and for which they are paying, it may be, five to ten per cent. more than if made at home. But over and above these, there are machines of various kinds of recent invention coming into extensive use (such as bread-baking machines) and these might surely be made in Ireland, if not under the patents of the inventors, at least such as would meet all practical purposes, and, unless we overrate Irish genius, such as might deserve to be protected.

In the laboratory of the chemist, the shop of the apothecary, and the various places where fire-clay, non-porous and other expensive clay goods are used, all are of English or Scotch make, and none the least better than might be made at home. In clay goods especially, and in fine clays particularly, Ireland ought really to be able to compete with any other country in the world.

For building purposes, in the repairs of houses, as glazing, plumbing, &c., there are large quantities of glass and earthenware goods used every day; besides chimney-pots, and, if they were not too dear, tiles would also be extensively employed. Now these "goods," and especially the tiles, could be made cheaper in Ireland than anywhere else, because there is not a class of tile for which the country has not a supply of clay. Encaustic tiles are now used in flooring dwelling-houses. They can be made of such patterns as bring them into use for ecclesiastical buildings, and as there is no mystery in making them, and as the very best of clay for the purpose may be found in the country, there is no reason why it should not be undertaken at once. On the farm, also, there is a large demand for clay "wares," and it ought to be a larger demand for iron "goods." But unless such works would promise a fair reward it could not be expected anyone would engage in them. Let, therefore, the cost of raw material in the various branches separately be put down, and the expense of works to utilise these be added, as also the labour; thus the cost of the several articles to be produced may be arrived at. Then let these costs be compared with the average current prices of the "goods," and it will be found that such profits stand over against each article as ought to induce the making of an effort. This will apply to every article of iron, glass, and earthenware alike. The profits, therefore, to be made on the home demand, taken at its lowest probable amount, are such as cannot be had for the use of money in England in like works.

CHAPTER XVI.

JOINT-STOCK COMPANIES.

THERE is a special reason just now for the further discussion of the principles of joint-stock companies, as applied to the development of Irish resources. On the data frequently stated already with regard to the differential profits on cotton and other textile manufactures, as between Ireland and Great Britain, it may truly be said that the extension of commerce can, in the present momentous crisis, be more safely entrusted to the operation of joint-stock companies than to private enterprise. Such organisations may engage profitably in the manufacture of cotton—including spinning, dyeing, bleaching, &c.; also, in all departments of the linen trade, and even in flax-growing. A company might undertake to supply seed to certain kinds of lands, and to take the produce at a given rate per ton—green, watered, or scutched—as a desperate remedy for the apathy of the farmers. In the manufacture, also, of woollens, silks, &c., and in the production of hardware, glass, and earthenware, capital so obtained may easily and safely be employed. There is really nothing that can be done by private enterprise which may not be done (probably not just so well, however) by a public company, whereas there are many things a public company may do well which private enterprise can never overtake.

Whether there be two persons or ten thousand, or whether they trade together under the designation of a firm or a joint-stock company, or whether the liability of each partner or shareholder be limited or unlimited, still it is the joint-stock principle that is resorted to; and, in all cases where it is fairly dealt by, it works well.

At present Great Britain has ten limited liability companies for every one in Ireland, although the natural proportions would be the reverse; because, whilst English capitalists were gaining their ends, now more easily attained by joint-stock companies they had to do it by other modes; whereas, it is only since the advent of limited liability that Ireland has been raising her head in modernised commerce. In all places the soundest opinions on commercial matters decide in favour of new undertakings being worked by this, it would seem the only principle by which the evils of undue competition may be met, and the interests of trade distributed to a proper extent. Dublin ought to be the centre of Irish joint-stock companies' operations. There is no reason, however, why many such may not be originated and carried on in the provinces. Yet, because persons from every part of the country may wish to take shares, there should always be a head office, at least, in the metropolis. A cotton manufacturing company, having its central board in Dublin, could choose sites for mills more easily, being alike near (as to cost of travelling on the average) to all the great water-power situations in the country.

If, for example, merchants in Galway, on the Shannon, in and near Westport, Ballina, Ballinrobe, in Cork, or other places, who have capital at command to the extent of from £20,000 to £50,000 each, were to take up cotton manufactures at their several places of abode, then at such particular spots joint-stock enterprise might not have so good a prospect as elsewhere. But so long as persons in such places neglect the materials at their command, it would be a hardship if such materials were allowed to lie waste, seeing that there are people in the counties of Down, Armagh, Antrim, and Derry, who have sums of money of from £20 to £2,000 ready to invest. But these persons, being merchants, tradespeople, or farmers, cannot leave their business to take advantage of the water-power of Connaught or Munster, for cotton or any other textile manufacture. These are no imaginary cases. It is known to all observers also that there are rich merchants in Galway, Limerick, Waterford, Westport, and other towns, where water-power is allowed to go to waste. Any one of these could begin the cotton, linen, woollen, silk, or other trade, on far more eligible terms than similar trades were commenced in Manchester, Preston, Bolton, Wigan, Stockport, or Staleybridge, Barnsley, Leeds, Bradford, Macclesfield, &c.

People having larger or smaller sums to invest in profitable undertakings are to be found in Dublin, Belfast, Newry, Ballymena, Londonderry, Omagh, Strabane, Letterkenny, as well as in many parts even of the province of Connaught, and through Munster. By strictly joint-stock companies, therefore, the work may be done. Certain wholesome conditions, however, should be understood as the basis of operation in all such enterprises. The first is, that the companies be under the Limited Liabilities Act; the second, that each such company be organised by *bona fide* shareholders, and every person paid according to the services rendered, but no one to get shares for patronage, or to be made director at the expense of the shareholders, because either of position in society or power in monetary circles; the third, that the owners of mill-site or other property should take shares in each company ; fourthly, that no one person be allowed to hold so many shares as would take the power of control from the general body of shareholders, or in any other way place the interests of such companies in the hands of an individual; fifthly, that each such company should call up a capital amply sufficient to carry on the trade, and keep the company free from debt, and out of the power of banks.

Any number of joint-stock companies conforming to these conditions could command support whether for cotton, flax, woollen, silk, glass, iron, or earthenware manufactures, irrespective of the seat of operations, whether in Galway, Limerick, Dublin, Cork, Derry, or elsewhere. The only object would be to get at the best materials, and to find out the place where they could be worked out most profitably.

If a landowner on whose property there is one or more good water-powers were to take a number of shares in a joint-stock company whose business it would be to erect a mill or mills at the water-power site, and work these in the manufacture of textile goods, such a landowner would at once secure good

tenants for his property, participate in the profits of the undertaking to the extent of his shares, and raise the value of every
acre of his property, without spending a sixpence. But there
are landlords who have not only sites, but mills built and wheels
erected. To each of these the long experience of several years
waiting "to let," and the loss of keeping a place idle, ought to
become a good and sufficient reason for looking after the getting
up of a company. There are not a few of this class in the
southern, and western, and the midland, and even in the northern
counties of Ireland. Others have the royalty of fisheries,
peat bog, quartz sand, or clay, and these have found also, by
experience, that but little comes of their royalties. The cost of
getting up a company (on the conditions before stated), and
which would in every other particular comply with the Act,
might be kept within 20s. per cent. on the paid-up, and probably
10s. on the nominal capital.

CHAPTER XVII.

SHIPBUILDING.

For textile manufactures Ireland naturally possesses motive-
power; and for iron, clay, and glassworks, materials on account of
which she claims special notice, from not only her own people,
but the manufacturers of Great Britain; but for shipbuilding her differential advantages are not of a like nature, and
yet shipbuilding is not the less feasible in Ireland than textile
or other manufactures. Nor is shipbuilding any less worthy
of the attention of capitalists, because it has an indirect rather
than a direct connection with industrial development. Nor is
its neglect any less likely to depress commerce, or its prosecution to promote the general interests of the country, because it
can only be done as profitably (in so far as some materials
affect it) in Ireland as in Great Britain; on the contrary, as
long as shipbuilding is disregarded, all other commercial enter-

prises must not only be more difficult, but less profitable. Therefore, as it is the interest of all who lay out money, or devote attention to any particular branch in commerce or manufactures, to encourage shipbuilding, every manufacturer, capitalist, banker, and merchant in the country are concerned in the matter.

But there are other and more weighty reasons why shipbuilding should be encouraged in Ireland, the most momentous of which is, that the employment of shipbuilding would gradually become a means of stopping emigration. This will appear from the fact that, as it is iron ships which are mostly made in the United Kingdom, the class of persons required in, and who could be easily trained for, such work, are the very class now going from us to Queensland, Canada, and other places. Bodily strength and mental vigour are as essential to the shipbuilding yard as in the new country.

There is much more than an accidental relation between the decline of shipbuilding in Ireland these thirty years, and the increase of emigration during the same period. The absence of a full set of trades (and there can be no full set without shipbuilding) is often the excuse for emigration. The introduction of a sufficient number of trades to meet the tastes and suit the talents of the Irish people, became more difficult as emigration became more extensive. The exodus of one class led to the necessity for others to follow. We do not mean, however, that to provide employment merely would be a sufficient reason for entering upon any course of trading or class of manufacture; but rather that one particular course being adopted, or a certain fabric or article of commerce made, the employment so given to one person leads to the employment of others of a family. But also, as has been illustrated in England, the opening of works for tradesmen and labourers attracts immigrants to the place of action just as much as offers of land. It was neither governmental injustice, nor the want of capital, nor the unsteadiness or the indolence of the Irish people, but the absence of the necessary workpeople, which sent shipbuilding out of the country. This want of people fitted for the several trades arose mostly from the fact that in the earlier stages of Irish emigration the picked men went first, and in the latter stages all who were able were obliged to go. Therefore none were left but those who were too weak, or unable to grapple with toil, or unwilling to change their abode, or those who had a stake in the country, which stake, whether in land or in trade, became less valuable as emigration became more the hope of the vigorous and well-educated. Nor can the loss entailed by emigration be estimated alone by the absence of persons fitted to take up trades, it also includes that of tens of thousands of small capitalists. These went to England or other places in the track of the talent and strength already gone. To all this may also be added the diversion of trade from Irish into English waters, which gave rise to Ireland drawing her supplies of teas, sugars, grain, wines, and other commodities, which she could have imported direct (and which were once so imported), by cross-Channel steamers.

But besides stopping emigration, fraught as it is with so many evils, the further prosecution of iron shipbuilding will attract immigrants.

Emigration was accepted as the remedy (about 1835) for the then evils of Ireland, which amounted to no more than the want of an outlet for talent and labour. But had industrial development arisen in 1835 as the remedy for Ireland's ills, she might have placed her manufactures side by side with those of England, and kept up her ports and harbours, and built ships, and carried on trade with any country on the globe.

England has been experimenting and achieving; Ireland has been looking on. The achievements of our sister may now be laid hold of, and the cost of experiments saved. We do not believe, however, that had textile manufactures been carried on in Ireland by steam, the advantages of holding a position in the market equal to that held now by England would have been so desirable as her present situation of being able to solve the problem of cheaper production. Besides, Ireland's days of industrial development being postponed, and her days of expansion having fallen in those of electric wires and submarine cables, as well as steamboats and iron ships, occasion is given for calling forth the talents of her people without that mere

drudgery which was undergone in the early stages of manufacturing and maritime enterprises.

A new generation has arisen in the last thirty years. These
are better educated. Ireland, with a population sufficient to
make a beginning, placed in circumstances to avail herself of
many hard-earned experiences of others which cost her nothing,
and having such an opportunity as she never had before, may
soon become more attractive as a field of labour, and more successful in commercial enterprise, than even Great Britain. No
one can visit the busy mart of Belfast, and see the general
activity of that place, without noticing the fact that, in a position less eligible than might be had in Dublin, or any other port
in the island, shipbuilding is carried on extensively, and, doubtless, profitably. Nor can the history of Belfast be thought of
without seeing that this branch arose out of the results of
enterprise in foreign trade. Shipbuilding was not the first thing
attempted ; nor is it likely to succeed anywhere by itself. The
class of persons necessary for the trade cover so wide an area of
talent and tastes, and also belong to so many families, that other
trades must be cultivated as a necessity of profit. Shipbuilding
cannot stand by itself; nor can any other manufacture flourish
long in Ireland, or compete successfully with the others, without this necessary addition. The Belfast people did not stop to
calculate the opposition of the Clyde. In this there is a lesson
to people in all parts of the country, and in regard to every
branch of manufacture. It is enough to know that the thing
will pay ; and that shipbuilding has paid, there can be little
doubt.

Fortunately, Belfast is not alone. Waterford has also made
a move. But these ought to be followed up by many other
ports in Ireland, and yet none would feel the loss of the trade
which others might gain.

The demand for iron ships is daily increasing, and with them
the means of enabling such places as Dublin, Cork, Londonderry, Galway, Limerick, to go into new trades, and of enabling
persons in such places to go into trade more extensively. We
consider iron shipbuilding no less, but rather much more, important than any of the trades we have already named.

Without more extensive shipbuilding, Ireland cannot get on. With an increase in the number of shipbuilding yards, the country may progress rapidly. Such would at once afford strong inducements to Irish people staying at home, and give attractions for persons to come to the country, to such an extent as would not merely render general manufacturing practicable, but also make that which is now doing more profitable. As an investment for money, this business offers a new field. Probably joint-stock companies might work it. Certainly partnerships are commonly resorted to in carrying it on. But, whether one way or the other, there is not a port in Ireland in which, for natural reasons, shipbuilding might not be pursued.

This is a trade in which Dublin ought to lead the way. It is one which would do more for the interests of the city than any other now available. It is not to be expected that any of the southern or western ports (Cork excepted) could compete with Dublin in iron shipbuilding. Dublin has at present just the class of population which would supply all the workmen, skilled and unskilled, in such a trade. As transatlantic and foreign trade demands more ships for Irish traffic, there is no reason why these ought to be built in the Clyde, Tyne, or Humber, more than in the Lagan, Liffey, Suir, Lee, Shannon, Corrib, or Foyle, except the absence of enterprise.

But all ships are not made of iron; there are still many timber vessels built. Ireland might also get a share of this trade. It is no more indigenous to Sunderland than it might be made to any other part of the United Kingdom, equally well situated for carrying it on. The advantages and disadvantages of the respective countries are rather, on the whole, in favour of Ireland *per se*, and greatly so as she is especially a rising country.

It seems difficult to say whether the demand for labour, or the labourer seeking it, is first presented. One thing, however, is clear, and if acted upon might soon lead to the solution of such a difficulty—and that is, that having so many labourers left even yet, it would not be unsafe for any one to invest money in the extension of shipbuilding.

In connection with shipbuilding there is obviously gross

neglect at several ports of the south, west, and north coasts. Into ports, including Londonderry and Cork, vessels are frequently driven by stress of weather. But, however, few shipmasters avail themselves of the shelter Irish ports offer, because, in case vessels require repairs there is either no way of getting them done, or, if done, it is at too great an expense. It is true vessels would not put in for repair unless they could not help it; but that is no reason why exorbitant prices should be levied. It is also true that many of those shipmasters dislike Irish ports. But it is quite possible to disabuse even the most prejudiced by giving good value, just as many an involuntary visitor to a shop has been made a steady customer by getting good treatment. The principle will apply in regard to ships, and we are perfectly satisfied, after the most careful inquiry, that many a valuable ship is pressed into the Channel at great risk, rather than make an Irish port if she requires repairs, because as much will be done in Liverpool or Birkenhead docks, or in the Thames, or other British port, for £70 as can be got at Cork or other Irish port for £100. This we have on the authority of American shipmasters, some of whom have tried both places more than once. The same testimony is borne by masters of British ships, themselves Irishmen. While such is a fact, it need not be wondered that the fewest possible vessels put in for repairs in Irish ports.

Ireland lost not her commercial prestige till she allowed emigration to flow and shipbuilding to decline. It was thus by neglecting the causes of decline she soon felt herself unable to overcome their effects. But a new era now gives promise of returning vigor, one index of which will be the regaining of lost ground in respect to shipbuilding, the first step in which has been taken and proved successful. It is a fact worthy of remark that the first steam-vessel which plied on the Thames (the Comet—not Bell's Clyde-built Comet) was invented and built in Ireland in 1805. It is also worthy of remark that in 1818 there were fifty steamers hailing from the Clyde, and but two from Cork, and none from any other Irish port; whilst from each of ten other rivers in Great Britain there were one to five, and from numerous places in America steam-vessels were being launched every year. The comparison up to 1851 was still more against Ireland. A reaction then set in, and at the present time it seems as if the former ground of perfect equality with Scotland and England, in regard to invention and execution in shipbuilding, may yet be fully resumed. This is probably one of the most suggestive subjects that could occupy the attention of the Irish people. Home-made ships would go far to give confidence in undertakings requiring shipping to carry out. Places for repairing in Irish ports would add greatly to the prospect of profits, and as we are on the eve, it may be, of a large foreign commerce, it becomes highly necessary every one interested in the country should reflect on the fact that if even Ireland can only make the ships she wants for her own trade, she may do so cheaper than bringing them from the Clyde or the Tyne ; and to make these would probably employ ten or twelve times as much talent, labour, and capital as is now so creditably invested and employed in Belfast.

Now that hopes of fishing by scientific laws, and not by random, to some reasonable extent, have dawned upon us, there is a trade open to fishing-vessels worthy of very strict attention ; and now that steam has been successfully adopted, and is likely to supersede sail (or rather to aid it), it becomes (if no other branch could be introduced) a matter of moment for the capitalists of Ireland to consider how far they could facilitate the development of fisheries by providing suitable steam-boats at once.

CHAPTER XVIII.

COTTON, FLAX, AND ALVA MARINA.

HAVING in the earlier chapters of this work touched on textile manufactures, we did not intend returning to flax or cotton; but the recent excitement about "cotton substitutes" offers so good an opportunity for again pressing the duty of immediate action in cotton manufactures, and the growth of flax to an extent never attempted as yet in Ireland, we have felt it right to re-open the subject. There are new reasons why those who have mill-sites with water-power to let ought to show signs of life, and offer inducements to capitalists. These, along with manufacturers and merchants, ought to take time by the fore-lock, and consider what arrangements might be made for an immediate expansion of the linen trade. It becomes still more imperative, in prospect of the future of cotton supply, as things now stand, that cheaper production should be taken advantage of, as a means of saving the trade from collapse. Agriculturists should also bestir themselves as to flax-growing at once. It ought now to be known, for the sake of manufacturers and for the interests of agriculture, to what extent flax will be likely to be grown in Ireland next year. We have had many opportuni-ties of hearing objections against the extensive cultivation of flax in Ireland. These, various as they are, do not justify us in saying that 1,000,000 acres is too much for Ireland. In the worst of times the average difference would be £2 to £3 per acre in favour of flax, and for the next three or four years it might be £4 to £5 per acre. We have come back to this topic chiefly because farmers had already become alarmed at the prospect of cotton substitutes. It is even whispered in some quarters that flax-growing has got its death-blow by Mr. Harben's discovery. We have, however, been saved the painful task of producing evidence to prove that agriculturists have nothing to fear from the ultimate value of alva marina as a substitute for cotton, inasmuch as it has already been found to be too expensive to get the fibre from the plant. But had this not been the case, the expense of new machinery, and, even after that, the impro-babilities of the fabric wearing, would have kept that article in a place more likely to encourage than to curtail the growth of flax. The discovery, however, is of far greater value to textile manufacturers than if it had really been useful as a substitute for cotton. It has called public attention to marine plants, and the fact has come out incidentally that in these there are the materials for *flax manure*, which might both increase the growth and improve the quality of flax. Alva marina may yet become more valuable indirectly to textile manufactures than the most sanguine hopes of its discoverer ever expected it to become directly. If agriculturists in Ireland are warned by these suggestions, and if they seek for manure from the sea, increased strength in flax fibre may be got, by which even the finest yarns may be woven by machinery. Machinery has done more to increase the quantity and lessen the cost of cotton goods than either the cheapness or the abundance of supply of raw cotton. If seaweeds be taken advantage of, there can be no doubt that flax crops may be grown—especially in Connaught and Munster—much more profitably than grain. The utilisation of the treasures of the sea may be made one chief means of raising Ireland to her national position in manufactures, and of increasing her commerce to its natural extent. But negligence must give place to activity. Those floating fears, founded on fallacies which have proved so fatal to many an effort, must first of all be dislodged. Considerable progress has been made in the linen trade since 1851. Finer yarns have been made than were ever made before, and machinery is being more introduced every year. In 1851 the finest yarns were 760 leas; in 1862, 1,000 leas have been produced. The former grist is 130 miles to the pound of flax, and the latter 170. The great desideratum is yet to be gained for linen as a competitor with cotton—namely, a more full adoption of machinery, to the consequent lessening of price and increase of quantity, so as to bring linen goods more. into use in domestic and other arrangements in which textiles are employed. Whatever may be the changes undergone in this cotton crisis in the relative value of linen and cotton goods, provided flax is grown extensively, and machinery introduced as fully in flax as it is in cotton (and this is practicable), the end will be that each will find its natural price, and both be used extensively.

There is no reason to suppose that flax could ever supplant cotton. Any efforts, therefore, to raise a flax-growing or linen-wearing or manufacturing mania is to be avoided, and ought to be discountenanced. So long as cotton ruled from 5d. to 9d. per lb. (mean 7d.), linen manufacturers, who have had a good deal to contend with as to supply of raw material, did not feel justified in buying expensive machinery to put linens in compe-

tition with calicoes. But now that there is no reasonable prospect of cotton ruling below 10d. to 14d. (mean 12d.), and that
flax may be sold at £55 to £75 per ton of 2,240lbs. (mean £65),
there is no reason for further delay in any necessary expense
for the purpose of raising the linen trade to its natural position.

The sets of linen most likely to displace calicoes are from
800 to 15 (light), and from 1000 to 1600 (strong). Those
coarser or finer will not be greatly affected, whether people take
to linen more and cotton less, or not. Tablecloths, diapers,
shirtings, handkerchiefs, have a trade on their merits; and the
only way these can be affected by cotton goods is that the price
of cotton will be too high to justify attempts to supplant linen
goods of those kinds. Flax and cotton may also come in contact in drills, ducks, brown hollands, blacks, and other dyed
goods; and especially for hot climates linen goods might be
made so cheaply as to give a very decided preference, and secure
a very large sale.

Nevertheless, cotton has its special claims also, which can
never be safely overlooked, and we still believe that, next to the
activity necessary for the special progress of flax manufacture,
the people of this country ought to look after cotton. It will not
do to depend altogether on linens. There must be an assortment, whether we sell at home or supply foreign orders direct
from our own ports. If we do not meet the demands of the
home trade fully, and go in for the foreign trade direct, there
will still be a serious drawback on the profits of every yard of
linen made in the country. To avoid all such drawbacks we
must cultivate cotton manufactures. There will be no scarcity
of cotton by-and-bye. The only effect of recent changes likely
to abide is the rise in price, which, whatever may be the special
circumstances to keep up prices for a few years, is only likely to
be such as will allow flax a fair field. But now is the time for
action in Ireland. The work to be done at home demands the
most ample preparation, and will require all our people. Every
acre of flax grown will employ three times as many labourers as
any acre of either oats or wheat. More scutching-mills have to
be erected. Spinning-mills and weaving factories have also to be
erected (even after all those now idle are set to work). Machinery

has also to be made, and some of it invented probably. Then there are hands wanted for the spindles and the looms, and bleachers, lappers, packers, clerks, salesmen, and all the necessary employees in marketing the goods, which demand on labour, although irrelevant to the subject in other respects, suggests the importance of using as much machinery and as little manual labour as possible, for other reasons than to secure cheap production.

As to the alva marina, whichever of the plants of its family is most likely to yield the sort of manure required by flax, must not be neglected. But this part of the business will be in other hands. It will not be necessary for farmers or manufacturers to look after the manure. That will come as the *debris* after extracting iodine and other valuable salts from seaweeds. There is, however, scarcely a trace of anything valuable in alva marina, and probably it would be most useful as manure, especially if it could be reduced to a portable size, and increased in nitrogenous properties by the addition of fish or some azotised matter.

But some objectors will say that whilst all this is being done in Ireland, there may be movements going on India, England, America, or some other place, to defeat all that Ireland can possibly do. We say plainly such is impossible. We are aware that fancy has figured out monsters to deter the Irish people from taking advantage of many bygone opportunities. But now the most important opportunity that Ireland has ever had must either be embraced at once, or be added to the many other evidences now extant that, were it not for the carelessness of her own people, this country might have been even more prosperous than any part of Great Britain.

The very same events which have plunged thousands in England into the deepest poverty, have offered Ireland another opportunity to develop her resources of industry, and it will be a sad thing if those are not improved. If danger to flax culture is feared from India cotton, then, by looking the prospects of that case in the face, we find that many years must elapse before roads can be made, the difficulties of language overcome, and spindles of the British Isles supplied. But the idea may be indulged that our new colonies will send us supplies of cotton, and so render the exertion of farmers to grow flax unnecessary, or to justify the indolence of manufacturers in regard to water-power and machine labour.

It seems a pity such objectors should not only reflect a little, as it would only require a moment's consideration for the most unreasoning to rid their minds from all fear of danger from these quarters. If, however, the people would cease to emigrate, now that it is a loss to do so even to the most attractive regions, those new countries might have some cause to fear from Ireland's activity in manufactures, and good reason to look to some other quarter for colonisers. There is no bounty possible to be offered, and no prospects of gain within the range of reason, equal to compensate the Irish emigrant for the loss he suffers in leaving a land so well fitted for manufacturing operations, and whose resources are as strikingly rich as they are shamefully neglected. The soil, the seas, the rivers, the geographical position of the country offer Canadas, Californias, New Zealands, Queenslands, and (the late Utopia) New Grenadas, far better than those across the seas.

But, some may say, England will not allow Ireland to get the advantages of the increased demand for flax. It is quite true she will do her best to gain a share ; but a little reflection here also will set the mind right. To grow flax, English farmers must displace wheat, beans, and peas, more than oats, and to do so would require them to get 50s. per acre more for their produce than would pay the Irish farmer, who would displace oats chiefly. Besides, that climate, which gives such crops of wheat, beans, peas in England, will not suit flax so well as this Irish climate, which will not grow beans or peas at all, but will grow flax better than any climate on the globe. With equal science and like industry the Irish farmer may make at least £2 to £3 net profit per acre more than any English farmer can of his flax, taking the value of the crops displaced to grow it in the respective countries into account. Then, as to the United States of America, if even all were tranquil again, it would take years to get trade into its wonted channel. But, besides the groundlessness of all fears affecting present duties, there is a prospect of demand, in consequence of the stoppage of so many looms for so long a time in Lancashire, such as, if there were no other reason, ought to stir the Irish people to the greatest exertions in providing both linen and cotton goods for every market in the world.

It may be asked what is to be done? Our answer is—first, get all false notions out of the people's heads as to English opposition. To do this, facts must be stated and reiterated, and the press alone can do this part of the work. Irish farmers, like those in Suffolk, ought to meet and determine as to flax culture, and take measures to get seed even now. Every member of society ought to make it his business either to learn enough of the subject to enable him to give a sound opinion, or to leave the matter to those who know it thoroughly. This is all that can be done just now. This being done, however, would be all that is necessary to establish and circulate sound opinions, and to excite the people to earnest and timely industry in this, the most hopeful opportunity Ireland ever was blessed with.

CHAPTER XIX.

FLAX-FIELDS, FACTORIES, AND FALLACIES.

So many are the opportunities afforded to Ireland this year for industrial enterprise, by the increase of her manufacture and the improvement of her agriculture, that in May, June, or July, next year, the number of fallacies still existing may be reckoned in an inverse ratio to flax-fields scattered over the landscape, and the number of factories to be seen at our river sides. The perversion of facts and mere apathy have led to the comparative neglect of flax culture, as also to the slowness of our capitalists in availing themselves of the best opportunities any people ever had for the increase of cotton and flax manufactures.

But new facts are now pouring in upon public notice from

so many unexpected quarters, that if only these are used judiciously by the persons whose business it is to do so, their use may dissipate errors and prejudices, and establish the knowledge of the special suitability of Irish soils for fibrous plants. These facts, besides, suggest the peculiar richness of our shore in those manurial elements by which such plants are nourished. Still more, no one can fail to observe that current events confirm the already safe conclusion that Ireland, above all other countries, has before her a successful future in textile manufactures.

Amongst the many valuable truths incidentally brought to light by the cotton crisis is the fact that in English factories there are probably 20,000 to 30,000 thousand Irish operatives. It is not long since the statement of this fact was really looked upon as rash, and probably regarded also as a stretch of imagination to gain a point. But it is now before the world, and its operation will surely destroy the notion that Ireland need have any difficulty to fear in manufacturing progress from want of "hands." This is a specimen fact, however, and must be valued as much for its indirect importance as for the direct point to which it is relevant. It has already been associated with kindred considerations—viz., rents of factories in Ireland and England, respectively; also, with the respective cost of water-power and steam-power in cotton or flax mills. But now that it has come to us from an independent source, and has thereby been established, we must ask for its associated facts also equal belief from the Irish public.

It would be presumption to pretend to foresee, and absurd to attempt to prevent, the natural consequences of neglect as to supply of raw material for a manufacture; but it is neither impossible nor impracticable, by taking warning from the " cotton crisis " so to cultivate various sources of supply that the cutting off of one may not again lead to such evils as are now seen in Lancashire. To do this devolves specially upon the Irish people. They alone are in the circumstances exactly suitable for doing it without great loss of capital. Such reforms as are necessary cannot be made in Lancashire, except at an immense loss of machinery. Thus our capitalists and merchants are ill-advised if they neglect to erect cotton factories at all places where water-power is available. It will also be necessary to introduce into factories the greatest possible number of machines, so as to reduce mere manual labour to its minimum, and bring up mechanical labour to its maximum. To meet the pressure of the times, as well for the sake of individual interests as for the general good, it is also required that flax mills with every new improvement should be established. Nor should flax-growing fail to attract the attention of capitalists and merchants, as well as farmers. Its increase ought to be encouraged by all means short of one class interfering unwarrantably with the affairs of another. But, as this is more of an agricultural than a manufacturing question, it is well to approach it from another point. The affair of the moment in reference to textile manufactures is to increase the number of Irish cotton and flax mills ; woollen and silk will then be safe also.

There is one reason in favour of flax-growing which so far removes it from the agriculturist to the manufacturing side of the case ; and that is, that as Ireland can certainly do so, she ought to yield annually as much flax as would, in case of a temporary stoppage of foreign supplies, keep her spindles and looms going, and so prevent such a crisis as that now experienced in cotton.

The very same primary reasons now exist to urge Irishmen into increased linen and cotton manufactures as formerly influenced the people of England to go so largely into cotton—namely, a desire to gain money by investing money, and to get rich by honest industry. These were the objects aimed at in England, and are still quite commendable, because, as strictly business matters, they ought always to stand on a business basis.

There have been greater changes since the dawn of 1862, in regard to the relative positions of Ireland and England as to factories, than have occurred in the whole of these last thirty years. Now the Irish capitalists are not only in the field, but are in a better position than English. It has, however, been deliberately stated by many that the people of this country can never become manufacturers and merchants like the English. This fallacy has been often exposed. Those exposures are written on the face of every movement in every country to which

Irish people are driven by the power of fallacy, as opposed to active industrial exertions at home. But of all the phases of current Irish fallacies, that one is the most striking on which is founded both blame against Irish people for not doing what they never had a fair trial of, and of accusations against Englishmen (who do not, as a rule, interfere), for hindering Irish people to make a trial.

No such thing was ever known as any opposition worthy of the name from people out of this country against efforts for the extension of Irish trade and manufactures. But there have been well-deserved sneers cast at " small-scale" efforts in Ireland, which invited an early annihilation. But the fallacy, in both its forms, as to the impracticability of the Irish and the opposition of the English, still exists, and must be dislodged before any great success can be looked for in Ireland.

The subject of flax culture is a rare field for observing this monster fallacy at work. It is said, judging by the past, that flax culture can never extend over Ireland, because the people have not been used to it; as also that unless there are factories started in the south and west, like those of the north, people, for fear of not getting a market, will never grow flax. To induce people to grow flax who have long neglected it, would be no good reason for a flax-spinner choosing the west for his operations; nor would the fact that spinning-mills had been studded over the banks of the Shannon, the Corrib, or the Lee, be any reason *per se* why farmers ought to grow flax. It is the business of the flax-spinner to choose, all things being equal, the place where he can make most money ; it is the duty of the farmer to grow the crop that pays him best. If both act on these principles of profits, despising crude precedents, then, but never before, will the interests of the flax-field and the factory draw their respective owners together, or cause them to respect reciprocity of interests in their dealings with each other.

The operation of this natural principle has been greatly hindered by efforts made to do by other and unsuitable means what efforts to make money alone can accomplish. In the first days of zeal for Ireland's prosperity some rushed into schemes only fitted to make bad worse. Persons who saw the sluggishness of farmers in less favoured districts, supposed that they were in need of professional teaching; and even now the same description of empirics would propose the founding of chairs in the colleges, to communicate special knowledge to candidates for the clerical profession, so that they may become teachers of the farmers, and regenerators of the domestic, social, and sanitary habits of the people! It must be a dangerous fallacy in which such a ridiculous experiment could be founded.

But, above all other fallacious notions affecting flax-fields and factories in Ireland, that of the unsuitableness of joint-stock companies to begin the work is the most unfounded. It is true this mode of commercial operations has been abused. But it is also true that for the amount of money turned over by the aggregate number of such companies, confining the calculation to those engaged in trades which might have been carried on by private enterprise, there is less malpractice and more profit than in undertakings managed either by a firm or an individual. The very best answer that could be given to those who ask how are the farmers of Ireland to be induced to grow more flax, is to tell them to put their surplus funds together in a joint-stock company, buy seed and resell it, then buy the flax on foot, or watered and bleached, or scutched from the growers; and, whether the operations may be in Connaught, Munster, Leinster, or Ulster, such a company could make money merely as traders in flaxseed and flax, and the fact of the incorporation of such a company would be a guarantee to the farmers of a market at all times for the flax crops.

CHAPTER XX.

FACTS, FUEL, AND FACTORY OPERATIVES.

MANY of the facts daily brought to light by the progress of the International Exhibition ought to be both interesting and ominous to such of the people of Great Britain and Ireland who

are engaged in, or dependent upon, textile manufactures for
their support. From the awards upon goods manufactured in
the places where water-power is used almost to the exclusion of
steam-power, there is one fact always apparent—namely, that
great and rapid improvements are being made in all those
places. In France, as also in Russia, and even in Austria, the
progress, since 1851, is full of warning to the British manufacturer; whilst the efforts towards cotton-dyeing, &c., as well as
spinning and weaving, in Prussia, Spain, Holland, and Belgium,
prove that British spindles, looms, dye-works, &c., are not the
only sources for the supply of piece goods and yarns to the
markets of Europe. In India, America, and wherever else
attempts have been made in textile manufactures, it appears
that the success is even more rapid than was expected. Therefore it becomes necessary now for those who have had the
prestige in these goods to bestir themselves, so that they may
keep their ground, at least so far as to secure, by fair competition, a full share of the textile goods trade.

To the people of this country it must be somewhat de-
pressing to notice the small part Ireland has in the exhibits of
the Exhibition. Doubtless the comparisons to which the goods
of foreigners have been subjected this year have suggested
greater improvements, and the opportunities they have had of
seeing machinery at work, and the facilities afforded for its
transmission in a short period to any part of the world, will
have great effect in stirring them up to activity. Probably
about four-fifths of all the cotton manufactured in every part
of the world is manufactured in Europe. Of this—or, say,
1,500,000,000 lbs.—1,200,000,000 lbs. is manufactured in the
United Kingdom, and only about one-third of the power for all
this is water-power. In woollen factories the proportions of
steam-power and water-power are about one-half of each; in
worsted factories only one-tenth is water-power; in flax only
one-twelfth; in silk one-sixth; and, on the whole, the proportions are about one-fifth water-power and four-fifths steam-
power, which, taken with the fact that, in India, Africa, America,
and all over Europe, water-power is almost the only motive
power used, suggests the possibility of saving between two and

three millions per annum in motive power alone. All special work in this way, however, devolves on Irish, not English manufacturers.

If the British manufacturer fails to improve present opportunities to place himself on as good a footing as the foreigner is trying to gain, by every fresh effort in machinery, the consequences may be as bad as those which have followed the neglect of Indian and other cotton-fields, so long as enough could be had from America, but it can form no excuse for Irish apathy in this the hour of Ireland's opportunity. It may seem a hardship to contemplate losses by the abandonment of sites which have become centres of operation, but in this case the first loss will be the best, and so sure as the expectations excited by the cheap cotton of 1845 have vanished, and as certain as all the fortunes made in that year could not avail (if offered now for the purpose) to beat back the consequences of past carelessness as to cotton supply, with like certainty will the time come, if present opportunities in regard to water-power are not improved, in which the value of one-half the whole machinery of cotton in Great Britain would fail to compensate for the losses which must then be suffered.

Few recent writers seem to notice the palpable fact that this country is in a transition state. The tide of emigration has ceased to flow so rapidly. The confidence of aspiring, young, able-bodied Irishmen is less in land and more in mechanical trades. The people generally are beginning to consider the propriety of more extensive manufacturing. On the whole, the appearance on every side is either like that of old ruins being pulled down, or of foundations being cut out for new erections ; and these remarks apply to both agricultural and commercial operations. In these circumstances it is no wonder that persons who take a flying glance at the country, and who only judge by what they see in one place, it may be, are led to make remarks which lead others astray. To know Ireland, a person must live in it, and see every part of it. The fact is, the country is flourishing on the whole, although some interests are declining. The people are awakening to action in every direction. So far from the country being in a worse state than formerly, all statistics to the contrary notwithstanding, the people of Ireland are in the very best condition in which those who were so long apathetic could be.

But, to return to cotton. The time is coming, and we believe is near at hand, when the subject of economy in fuel must be dealt with. Railways are extending, and these are consuming now nearly 2,000,000 tons of coal annually. Steam navigation, so rapidly on the increase, is also absorbing ten millions of tons every year. The export trade is growing, and, though geologists have generously given the coal-fields a lease of four thousand years (there are no reliable data for this), yet, the raising of 80,000,000 of tons annually must tell, if not sensibly on the area of coal formations, yet certainly on the ideas of coal proprietors, who, like all others, have the more power to regulate prices the greater the number of interests dependent on supply, although the question of so many more or less tons of coal being burned as the turning of textile fabric machinery might require, is not a point of much moment *per se*.

But, suppose changes come by which it would be necessary to raise the price of coal (and there are indications of such even now), so certainly as any large rise in price, however justifiable, would be attempted, it would be resisted, and a crisis more or less momentous would certainly ensue. In such a case, therefore, it would be a matter of no small importance to have the machinery employing 4,000,000 of hands, as free from risk as possible.

It must be within the recollection of all that the blindfold policy of depending on America so much for cotton was not indulged in without warnings being faithfully given. It was argued there that 39,000,000 more acres could be added to the cotton-fields, and 800,000,000 bales be produced if necessary. In the same direction it was urged that West India, South Africa, and even new places only just found out, could be made to grow cotton for Liverpool market, and sell it at 4½d. to 5d. per lb. All this was theoretically true, but failed under the simple yet exacting demands of commerce, which, in reducing ideas to practice, allows nothing to go on but what makes profits. All things, therefore, resolve themselves in commerce to the profit and loss account at last, so that whatever is thought of ought at first to be tested by the probable figure it would show on that page of the books of commerce. Everything may be safely attempted that will pay, but nothing can be touched without loss which is founded on a violation of the rules of book-keeping. There can be no better reasons assigned now for further disregard of the saving of fuel by using more water-power than were presented in favour of letting " well alone," so long as the United States of America could export to European ports over 2,000,000 bales, and keep for the spindles in the Northern States 1,000,000 bales besides, every year.

There is still another consideration in favour of water-power with manufacturers in Ireland—namely, the condition of the operatives. Whether these be many or few, their circumstances must be improved. Water-power mills may be erected in every part of the country. These being scattered over so large an area will preclude the possibility of their ever becoming the occasion of any such evils as those which have arisen from the congregating into one large town (or into a narrow district for sake of coal) of persons, many of whom must have left their homes in order to work in the mills. Besides, the very situation of water-power mills, in a sanitary as well as a social point of view, would be favourable to the operatives. There might be mills on the Lee, the Corrib, the Suir, the Bann, the Shannon, and the Liffey, to and from each of which, by the adoption of the tramway (to shunt goods in and out of mills), and the " lighter" to carry them alongside the ships at Dublin, Belfast, Cork, Galway, Limerick, &c., the cost of getting these goods to the seaboard, or that of getting supplies to home markets, could not be reckoned at any rate per cent. on profits or price per piece on goods. It would, therefore, be profitable as it would be practicable to scatter the factories of the United Kingdom over the river courses of Ireland. This is certainly one of the most available, if not the only means at command for the cure of social and sanitary evils in the factory system.

At present, on the surface of those places in the United Kingdom where coal is most plentiful, commerce and manufactures are most prosperous. But it may become equally true in less than thirty years that where the best water-powers are, there trade flows most liberally. The reason why trade is now best in coal districts, and ought to be best in future near large rivers, is identical, namely, because at such places profits may be made most rapidly, and trade will most concentrate. The expansion of the cotton trade in England has led to the extinction of several lovely villages, which are now lost in large towns. The smoke of factory chimneys has made many a landscape cloudy which was fifty years ago bright and clear. The very opposite would be the effect of the water-power system. Its adoption would become the occasion of creating new villages rather than monster towns and cities, and of ruralizing once more, and on a permanent basis, the manufacture of flax, wool, silk, and cotton, in Ireland.

CHAPTER XXI.

BUILDING MATERIALS AND MARBLE.

If the conclusions of many whose opportunities for observation are too limited to justify the expression of an opinion were accepted, it would appear as if Ireland were retrograding rapidly. In such a case building materials would be valueless, in a commercial sense, and marble would be worth no more than the commonest flagstones. But, on the contrary, seeing that there is evidence of steady, healthy, and hopeful progress in the agriculture, manufactures, and commerce of Ireland, there can be no more important work suggested for all persons interested in the welfare of the country than a careful inquiry into the extent, localities, and probable cost of building-stones, bricks, lime for mortar, &c., ornamental stones, flags for flooring, roofing slates, marbles, and whatever else may be required in erecting or ornamenting factories, warehouses, dwelling-houses, business establishments, public buildings, &c. Such is the nature, moreover, of the improvements now going forward, that

Irish progress implies pulling down old erections and the building up of new ones to an extent never before known in this or, probably, any other country. Agricultural reforms imply the necessity of a new style of farm buildings. So soon as manufacturers, to any reasonable extent, are engaged in the introduction of more dwelling-houses, more warehouses, more offices for banks as well as mills and factories, where the like had never been before, will so change the aspect of the country, as that towns will be raised up where there are only villages now, and villages will rise up in places at present desolate—all which imply the use of stones, brick, slates, marble, glass, timber, and other materials, some of which are to be had at home, and some must be imported.

Those who judge hastily fail to see the most important points of improvement in Ireland's condition. It must be admitted that all changes in Ireland within the last ten to thirty years, but especially those of this year, 1862, have, on the whole, tended to permanent and substantial improvement. In these circumstances the value of Irish building materials has increased. If, however, stone, slates, chimney-pots, &c., must be still imported from England or Scotland, now is the time to settle the matter so that persons making calculations for building may do so safely. But prosperity has set in even already so far as to render the present mode of getting building materials so anomalous as that to look forward to a future of any degree of building activity, with the sources for supply of materials in their existing condition, would be to begin to build, indeed, without counting the cost. For some Irish buildings English and Scotch stone and marble are imported. None ought to be imported, provided the Irish stone and marble are as good, and can be had as cheap, as the English and Scotch. That ours are as good, has long ago been settled. Although theoretical geology has not been cultivated so much in this country as in Great Britain, yet it has got so far as to supply evidence on which anyone may safely believe the supply of granite, sandstone, slatestone, marble, roofing slates, clay for bricks, tiles, &c., to be as good and as plentiful in Ireland as in Scotland or England. The difference, however, between Ireland and Great Britain in respect to the

use made of these materials, may be traced largely to the fact
that the study of geology became suggestive in the latter country
of enterprise, but here, for so far, it has been used for little more
than to mark the distinction between the learned and the illiterate.
Amongst those adventurous people who are ready to lay hold of
every fact, and press it into some active service, the science of
geology has been of the greatest possible aid to commerce.
But so long as apathy exists, and that the valuable stores of
Irish building materials are kept from public gaze, it will matter
little whether or not our young men follow the example of their
fellows in Great Britain, and make geology a branch of an ordinary education.

Commercial speculators regard not the mere existence of a
commodity, but rather its extent and value, as also the facilities
which exist or may be introduced for its transmission to places
where it may be required. In this respect, then, the comparison
of British and Irish building materials shows decidedly in favor
of Irish. Wicklow abounds in granite. Some of this has been,
and is being, sent to Scotland and also to England. But it
might be made still more productive in a commercial sense.
There are granites also in Carlow, and some in the county of
Dublin. But the red granite of Galway much surpasses in
commercial value all those which are dull and lustreless. These,
however, are no less useful for building, and may yet be required
largely for making breakwaters and docks in Irish bays and
harbours. Both Mayo and Tyrone possess a little granite.
Donegal has a great deal. Down abounds with the best of
granites, some kinds of which, to a small extent, are now worked
at Newry, and may be more so when the railway is made, *via*
Rathfriland, to Downpatrick, on the N.N.W. side of which
town, extending a distance of two miles, there are several specimens of both red and grey granite. If any one wants red granite,
however, it is actually easier to get it from Aberdeen than in any
part of Ireland, even though the Irish specimens are better than
the Scotch. The fact is that, some years ago, large quantities
of granite were imported from Scotland to Dundrum bay at a
less cost than they could have been got in Ireland, simply because in the quarrying and trading of granite, skill, capital, in-

dustry, and enterprise are invested in Scotland, whilst none of all these are applied in Ireland, even though the granite would be more easily quarried, and quite as near to the seaboard, as that of Scotland. Here is a field for enterprise, and one on the cultivation of which money might be made at the rate of eight to ten per cent. But, bad as the granite case is, that of roofing slates is worse. In contravention of every principle of sound commerce, and to the loss of thousands of pounds per annum, the people of Ireland import slates from Wales, and allow their own quarries at Killaloe, on the Shannon; Valentia, near the seaboard; Glandore, county of Cork; and Ashford Bridge, Wicklow, to be comparatively neglected. These have never been worked deep enough. There is reason to believe that at like depths to which Welsh slates have been worked Irish quarries would yield as good qualities as Welsh. Those already taken out therefore are not fair specimens. We do admit that the cheapest and the best should always be purchased; but we do not admit that Irish slate quarries have yet got a fair trial. The demand has not been made for these slates to the extent to justify that outlay of capital necessary to reduce the cost of working to a minimum rate. But the fact is, the reason why the Welsh slates can be supplied cheaper than the Irish is that those who began to work the former did so on a scale to enable them to command a trade; and if Irish capitalists do likewise, they will find that no Welsh slates can compete with Irish. In default of this extent of operation it would be more for the interests of commerce not to work Irish slates at all.

Bricks come next. It is true of every part of Ireland that bricks can be made, and the range, from the highest cost to the lowest, may be taken at from 10s. to 30s. per 1,000, the maximum of which ought to be greatly lessened in case home fuel, and especially coal, were more extensively used. No country could be better off for the means of making bricks than is Ireland. In Donegal, Fermanagh, King's County, and Galway especially, bricks may be made very cheaply. This is suggestive of a saving to those who might wish to erect factories in the West.

The native stone of each county in Ireland is so abundant,

that building to any extent may be cheaply carried on in every part of the country. This is certainly a matter of more moment than the export of granite. Kerry, Cork, King's and Queen's Counties, Meath, Westmeath, Roscommon, Galway, have the best of limestones, which make the most beautiful of buildings, as may be seen in each of these counties. Waterford, Wexford, Carlow, Kilkenny, Tipperary, Limerick, Clare, Kildare, Dublin, Longford, Mayo, Sligo, Leitrim, have sandstone, freestone, lime, and granite, all of which offer supplies for building purposes not to be surpassed anywhere. Antrim, Cavan, Derry, and Tyrone, have sandstones of the finest kinds, which might be worked so as to be sold cheaper than any stones of like quality in any part of Great Britain. Bringing this class of stone to Derry, as was done in 1859 for the Magee College, was far more absurd than taking coals to Newcastle. Monaghan has also both lime and sandstone. Louth, Fermanagh, Armagh, and Donegal have each a variety of the best specimens of those stones most useful for building purposes. Down has both granite and sandstone. Kildare has also a good variety, so that every county in Ireland possesses within itself, and in rich abundance, all the materials for building, except timber. In all respects, therefore, Ireland is better off for building materials than England or Scotland. Besides, she has an equal advantage in nearness to America for timber, as in richness of supply of stones. Lime is also abundant and cheap in this country, and it need scarcely be said that fresh water sand is easily obtained.

The relevance of these facts, however, bears chiefly on the future. Its only importance at present resolves itself into the question, why do the merchants of Ireland, or rather why do the builders in Ireland, not buy Irish materials? We need not, perhaps, say that we do not wish to promote an Irish building material mania; but we do really suggest to all who are interested in the welfare of the country, that a matter so important as the rejection of better and cheaper materials for worse and dearer ought to be inquired into and explained. At present we know of no explanation which would not reflect disagreeably on the commercial skill of Irishmen. Probably there is some reason in the fact, it is easier to get credit from dealers in

English than from those who deal in Irish materials; but if this
be the case, it only proves that the want of capital in the business is the only barrier in the way of its prosperity. We are
happy to say people are beginning to consider these matters
more. It must not, however, be lost sight of that, before the
difference in the cost of Irish and English building materials, as
compared with English and Scotch be felt, there must be more
building, and before that there must be increased trade as the
result of extension in manufactures.

The marbles of Ireland are a valuable, and, we regret to say, an
unworked treasure. These, like some valuable ornamental stones
in Ireland, equal to, though not exactly like, Portland, may be
comparatively disregarded until an increased circulation of money
leads to the use of more ornamental works in the inside and on
the exterior of their dwellings. At the same time, it may be
well to say that, for all kinds of marble, except pure white, Ireland cannot be surpassed in any part of Europe. But there is
pure white marble near Dunlewey, in Donegal. There is also a
specimen at the same place, which, except for its approaching
a dove colour, might compare with the marbles of Carrara. It
is a mere question of taste, however, and one on which there
ought to be room for difference of opinion as to whether the
colour of the Irish were not more chaste and elegant than that
of the Italian marble. But leaving this point, to which we only
desire to glance, there can be no controversy about the black
marbles of Kilkenny, Cork, Galway, &c., the green and
variegated of Connemara, and the several kinds of so many
shades to be found in Armagh and other parts of the country.
Neither England nor Scotland can compete with Ireland for
marbles. It is probably one of the most cheering signs of the
day for Ireland that notice is being taken of her marbles, and it
augurs well for the future that a trade is springing up between
several parts of Great Britain and this country for these valuable
resources of industry.

It is, however, to be regretted that so few Irish people make
the working of marble quarries a source of employment. If
capital were put into these to a sufficiently large extent, and that
along with the supply of chimney pieces, &c., marble were

worked into personal and household ornaments, there can be no question but with that enterprising, persevering, and pains-taking industry which are now necessary to all success, the marble business of Ireland might be made to occupy a high position as a branch of trade. The localities of marble are favourable for export. They are also suggestive of immediate action, because they are in the parts where trade is worst, and into which it is most desirable more means of employing labour, skilled and unskilled, ought to be introduced. The only hindrance in the way is the absence of enterprise. The blame of this falls solely on the Irish people, and to them alone ought anyone desiring to see the marbles of the country worked look for reform, and such aid only ought to be expected as would be consistent with a profitable investment of money, by which restriction any efforts made in the direction suggested are more likely to be facilitated than hindered.

CHAPTER XXII.

THE RELATIVE COMMERCIAL POSITIONS OF ENGLAND AND IRELAND IN 1862.

COMMERCE, wherever carried on efficiently, respects the relative value of all materials which are, or may be, manufactured into mercantile commodities; also the interests of every place where trade is prosecuted, and the rights of every person engaged in conducting it. It will be obvious to all who are accustomed to consider commercial subjects, that not only are these the indispensable requisites of mercantile success, but also that, in proportion as the trade of a country has been kept up to these requirements, it has flourished; and, in so far as it has been allowed to fall short of them, it has invariably declined. Great Britain, in the past period of her mercantile career, affords the best example of adherence to these desiderata. But few parts of the product of her soils, or of the materials to be had from her mines, have been allowed to lose their value by being sent to market in an unmanufactured state. Coals were used in making textile fabrics, the greater part of the raw material of which, and much of the food of the operatives engaged had to be imported. The factory demanded machinery, and machinery was made of iron, and that was also manufactured at home. In like manner other materials were made the most of, home materials being made to help one another in manufactures. Rivalry was usually founded upon merit, and rather helped than hindered progress. Whatever was to be made, imported, exported, or exposed for sale, found, on the whole, suitable places. The result was, as might be expected, that, not from philanthropic sentimentality, but for the sake of individual advantage, the merchant respected the interests of the retail vendor; both regarded themselves as in the same ship with the capitalist, and all agreed to make common cause with the skilled, and in most cases with the unskilled labourer.

Ireland, at the same time, afforded a striking example of the disregard of all those principles by which commerce is sustained, and those engaged in it stimulated to activity. Attempts at the manufacture of flax, cotton, or wool, on a large scale, and by the use of the best machinery in Great Britain, without taking advantage of coal for steam-power, would have been a violation of common sense. But to neglect water-power, and use steam-power for like purposes in Ireland, was ten times more absurd. Perhaps one of the grossest breaches of principle that can be found is to be seen in the fact that Irish capitalists and merchants did not, when tempted to do so by the success of their near neighbours, adopt the principle of using home material, and enter the lists as competitors for a share of that trade in textile fabrics for which Great Britain is so famous, and in which Ireland might have held a higher place to-day than any country in the world. But places in Ireland have suffered more than things, or even persons, by this mode of procedure. Persons emigrated, and attained in many cases eminence, and in all considerable success. The water-power is still available. and though it is a matter of deep regret that Irish fisheries, mines, &c., which ought to have been worked, have been so long neglected, yet these have not diminished in value, and may be worked up; but towns once prospering are now comparatively deserted, villages once joyful have altogether disappeared, and homesteads, in many places, been levelled to the ground, yet Ireland has not-sunk entirely, nor has Great Britain risen so high as not to be rivalled. The positions of the two peoples, we cannot too often repeat, have altered in 1862, and new responsibilities rest upon each. The English cannot preserve their cotton-trade unless they avail themselves of water-power. The Irish may now extend their flax trade, and add to it cotton, wool, and silk, so as to regain the ground they have lost. · Rivalry, to some extent, as a consequence of the enormous profits arising from cotton, has set in between towns in England, and as this sort of rivalry worked only evil for Ireland, so will it work no good for England. This fact, which is apparent to every observer, urges the transfer of cotton-spinning to Ireland, for reasons wholly British. In any case, however, Great Britain has more to gain than Ireland by the adoption of our water-power. This transfer of trade is also suggested by that adherence to principle from which commerce grants no release. It is as necessary for the English manufacturer to come to Ireland now, in order to make the most of his machinery, as it was some 30 years ago for him to adopt machinery in order to make the most of his capital.

But still more such a course is urged upon immediate attention as the only means of preserving the cotton trade, now that English-made machinery has been adopted by those who have the advantage of driving it by water-power at from one-sixth to one-fourth the expense attaching to steam-power. In Great Britain, there was little heard of the unsuitableness of London as the great commercial capital and the centre of monetary operations, so long as Liverpool was struggling into life, and Manchester was but beginning to feel the favourable effects of cotton spinning, Wigan the advantages of coal mining, and when Bolton, Stockport, &c., were only known as good market towns; whilst Staleybridge, and several small places now of cotton-spinning celebrity, were unknown. But latterly it seems as if the error of the Irish people were to be matched in England, which fact seems to be overlooked by those whose imaginations conjure up combinations in all places in England (where cotton is spun) for crushing every effort that might be made in Ireland. But Dublin is now rising into repute everywhere in Ireland, no combination of trades exists in England against Ireland, circumstances at home favour action, we come, therefore to the conclusion that those who say that nothing will ever be done in the cotton line in Ireland, are deciding against evidence.

Still more the advent of the Atlantic Company, if it served no other good purpose, has been the occasion of introducing a better spirit amongst the people of the different provinces, and the several ports, cities, and towns of each. The people of Dublin can now hear with pleasure of the prosperity of Cork, Waterford, or Galway, and hail the rapid strides of Belfast and Derry in commerce, as facts in which they have an interest of no inconsiderable value. This better state of things goes even so far that now when people in any part of Ireland hear of any town— Belfast, for example—rising in a particular manufacture, such as shipbuilding, for instance, they neither (as used to be the case) feel sorry that it is not their own place, nor are alarmed or excused from effort by the tidings. The feeling in Ireland is rather that all progress should be approved, and the industry which such prosperity implies be copied.

Whatever measure of prosperity Ireland has had of late is obviously because capitalists and others have paid more attention to home commerce; also, that the press has been more engaged in treating of industrial resources; still more, that greater industry and more self-reliance are applied in their development, but more particularly because it is a growing sentiment that persons in all ranks, and those of every place, are alike concerned in the material prosperity of all parts of the land, and in the maintenance of the dignity of the capital for sake of the interests of commerce. This can only be properly answered by the fullest employment of Dublin as the commercial and monetary centre of Irish trade, as London is now of British. It is happily even more apparent that Irish prosperity is founded on a better understanding of the real value of things and the relative interests of persons engaged in agricultural, manufacturing, and commercial labours respectively. The mining operations of this country have contributed more to her prosperity, even though these are only as yet in embryo, than all mere trading in goods made elsewhere ever did or ever could do. Besides the little that has been done in mining has created a demand for scientific and skilled labourers.

Tillage of the soil is, after all, the first concern for Ireland; but that this may be done in the best style, and its products sold at the highest rates, the example of Great Britain in putting foremost manufactures, must not be overlooked.

CHAPTER XXIII.

BANKING—LIFE AND FIRE ASSURANCE.

As it is not of banking generally, but merely of Irish banking, and of it in only some of its aspects, we are to treat, we cannot allude, as we might have done under other circumstances, to a question which seems to press on public attention—namely, the relative usefulness of privileged and non-privileged banks in commerce generally. Nor can we say more in regard to the principle of Irish banking than that, according to present appearances, free trade is likely to be much more fully extended to the business.

That banking operations have ever preceded, or at any time called forth, commercial activity, is one of the grossest errors which could be held in regard to the origin of banks. Every chapter in the history of Britain's commercial progress becomes testimony against this fallacious notion, which, we regret to find, has still its adherence even amongst the most prominent merchants, capitalists, and writers in this country, even though its withering effects on trade and commerce grow daily more apparent and distressing. Before any bank which has proved itself useful to trade and profitable to its shareholders was ever established, the necessity for its assistance to complete what had been begun, and not for its help to relieve from responsibility, had been fully proved by the extent to which trade had been carried. Nor is it any less evident that, if a second or a third bank be got up in a place merely to oppose a first, and not (as is almost always the case) to meet the growing wants of trade, such bank will not only prove a loss to its shareholders, but must also interfere, less or more, with the objects of banking and the interests of trade.

The first part of these statements has been exemplified in every class of bank, and in all banks of every class which have entered on the legitimate mission of becoming at one time the custodians of surplus funds and the medium of their investment, and at another the repository of aid at call in money, or in negotiable equivalents, according to the exigencies of trade. Nor have the illustrations been few of the fallacy of attempting to hinder the progress of sound banking by attempting impracticable or utopian schemes under the name of banks, whether of deposit, issue, or accommodation.

But the mission which not a few persons in Ireland expect a bank to fulfil is something like that of a substitute for enterprise—a sheet-anchor for such distress in trade as negligence or indolence may bring about; and because of this expectation obtruding itself in the face of attempts to extend banking up to existing necessities, and not the unwillingness of bankers to accommodate traders, the banking of Ireland is as it is to-day— far below the requirements of trade. There is a notion, however, in the minds of persons in most places south and west of Dublin that they ought to look to other quarters, and not to depend on their own exertions for the initiations of a better state of banking. This is a dangerous delusion. There must be self-reliance as a basis of banking, and the way self-reliance may the most effectually impress itself in this case is by creating the necessity for more banks. The people of Ulster have proved themselves equal to the task of banking, and so have our cousins across the Channel southwards, as well as the celebrated North Britons; and, on the principle that those who have most experience are best fitted to deal with difficult cases, we believe the earlier stages of the extension of banking south and west might be safer in the hands of English, Scotch, and Ulster people, than in those of Munster or Connaught. But we by no means fall in with the idea of some, that what suits in Scotland or in England, or even in Ulster, would, therefore, suit in the south and west. But little good can ever come in this country from the mere operation of agencies of a bank whose management is in places of character so widely different from that of Ireland as to forbid the possibility of applying any common rule to good purpose, except by a miracle. There is plenty of money in Ireland, and in no part is it more abundant than west of the Shannon. That which is wanted, therefore, is to draw it out and make it serve the ends of commerce. Such would be the first duty of the bankers in those parts. The next is to administer the aids of banking accommodation so as to obviate the necessity for the small trader going to the usurer. Perhaps it might also embrace an effort to keep him from the loan fund. How far the loan fund system has aided the poorer classes, is a question on which we hesitate in stating the only opinion we feel justified, from all we can learn, in holding—which is, that they are more an evil than a good. But we must allow that the intentions of those who promoted these as aids for small traders, and helps for the poorer small holders of land, have been answered, and that the evils which attach to their working are small as compared with the " three prices" of the mealmonger, and the " 40 per cent. and two solvent sureties" of the usurer, to which these institutions have given almost the death-blow in the lower stratum of trading.

But, this stage having been gained, we see no reason why the naturally elastic principles of banking could not be so systematised as to meet the wants of all traders and farmers who would be entitled to monetary accommodation because of their enterprise, industry, and good character, over and above their solvency, which ever must be a condition of advances from banker to customer.

But whatever is done, whether in eradicating those evils which have been already mitigated, or in meeting the wants of traders in Ireland, those whose interests are favourably considered must never expect to be released from personal responsibilities, nor to have aid extended beyond that which is fairly earned by individual merit, and sufficiently secured in the ordinary sense of the term.

There is, probably, no department of business in which reform is so called for, and so slowly adopted in this country, as in banking. We hesitate not to say, judging by the common mode of administration, making all necessary allowances for the difficulties bankers have to contend with, that so soon as life takes possession of the traders of Ireland, there will be as much progress made in two years as will prove the present Dublin system on the average a whole century behind. Even in the little trade now done, the banks of the metropolis do not take hold of foreign exchanges on Ireland, but allow them to go to London; nor have they ever become as they ought to be, the natural reliance of joint-stock company and other enterprises. These, too, are sent to London, or Manchester, or Liverpool, or anywhere, provided it is out of Ireland.

The establishment of the Dublin Branch of the Ulster Bank, and the opening of the Union Bank in Dublin, are signs of improvement. These, however valuable, are not alone. Co-operation on sound principles, and such as has been so successful in England, is also beginning to arise in Dublin and elsewhere. The thraldom of trade under the yoke of exorbitant interest is somewhat relaxed, and the state of the country is, on the whole, so far satisfactory that persons are beginning to find out the errors of the past, and, what is still better, are rather inclined to active measures for reform than to sorrowing after what is gone, or sighing over what cannot be undone. These general outlines of the state of the country in regard to expectations from banking, as well as the fact that thousands of pounds still rust in "corners of walls" and "old stockings," indicate a probable extent of business well calculated to stir our capitalists into such activity as would first clear rubbish out of the way of banking, and thereafter (the way being cleared) push forward operations to an extent not only equal to the wants of existing trade, but such as would assist traders to expand their operations up to the capabilities of the soil, the mines, and the manufacturing powers of the country.

Banking seems a far more formidable thing to a Connaught trader than to one in Ulster—which fact accounts largely for the greater tendency in the west to hoard rather than to use money. It is also observable that, in proportion as people are familiarised with a bank, in like proportion do they become rightly affected by the value of money. Therefore, in addition to the making of liberal dividends for their shareholders on money which would otherwise have remained latent, and over and above the assistance given in the creation of wealth, the Belfast banks have also morally done much good to the people in every town in which they have opened branches, by showing them that money is only good for its use, and that its use is within the reach of all who are likely to use it wisely—refused to none but those who are likely, if allowed the chance, to be guilty of abusing it.

Although secondary influences like these have not followed the track of banks in the south and west, yet the reason is not so much the better skill and more liberal views of the managers of the several Belfast companies, but in the character of the traders in the places where the others had to operate. In Belfast there is a greater tendency to become manufacturers and merchants so soon as a little money is made, than is seen in Dublin, Cork, or Galway. There the desire is to become money-lenders, and oppose, or rather so far defeat, the ends of banking. Those on whose struggles the iron grasp of the usurer has lain the heaviest, often, after getting release, become the severest on others who may be their victims. The failures of those traders on capital borrowed from successful shop-keepers, whose object in lending at the best is to get exorbitant "use," and whose interest it may be to crush the trader by exacting payment on short notice, are now so often exposed to the commercial world, that such cases, for mere shame, seldom come before the light. Yet many such there are, and all hang on the progress of trade like a millstone on a man's neck. To these cases and to their consequences may be traced many of the special difficulties traders of this country have to undergo; evils which those of England, Scotland, or any other commercial country know little. Some have, however, traced special difficulties falsely to the desire of merchants to exact larger profits from an Irish than a British shopkeeper, which is not the case. These and many other evils attributed in trading circles to the want of a good system, or the existence of a bad system of banking, have nothing to do with banking matters at all. Some cry out that so long as the legislature does not interfere in this, that, or the other case of bank administration, trade cannot flourish. The fact is, however, that no power can affect the traders of Ireland through banks beneficially, but those which operate through plans every one of whose primary objects must be something personally profitable.

Private banks are just as capable of doing all the offices of banking as chartered or joint-stock company's banks. Those under limited liability are just as well fitted for all the purposes of business as those whose shareholders are not so protected. That in which lie the safety, efficiency, and advantage to traders of a bank cannot be affected by any accidental form of organisation, but must either fail in or fulfil its mission in proportion to management. Management, to be good, must not be according to formulæ so much as in keeping with principle; nor ought it to be so much by rule as by a desire to meet in all particular instances the speciality of each individual case. In so far as the traders of this country cease to expect banks to become merely fountains of supply, whose source it is not their business to keep up, in that proportion will they be likely to attract capitalists to embark in the business.

Savings banks in Ireland have had considerable development, and have proved useful to no inconsiderable extent. But such are the peculiarities of Ireland's present condition, that most persons of provident habits could do better with money than merely lay it up in safe keeping. That which comes most into immediate contact with savings banks, both in a monetary and moral sense, is life assurance. The savings bank has got more justice in Ireland than the assurance company, although the latter is calculated to do much more good to the community through the individual, than the other can do to the individual whose money is safe under the protection of government surveillance. The people generally understand the simple fact, that by putting a given sum at stated periods into the savings banks, they can rely on getting it again with some additions — almost at call; whereas, but few will listen to the claims of life assurance, just because these, unlike the others, do not come home to them by knowledge arrived at by intuition. Had the assurance companies of Ireland taken the trouble to instruct the public mind in the principles of life assurance, an anomalous numerical relation of savings banks deposits in comparison with life assurance policies, would not now, as it does, exist in Ireland. There are sad errors on these matters afloat in all parts of this country. Such erroneous notions often keep persons from assuring whilst in good health, and consequently led often to the collapse of business which may have been begun and carried on to a considerable extent, and which would have been completed if either of the original promoters had lived to finish it, or had left enough by policies of assurance to enable others to do so. This is not even so bad a case, however, as that of farmers who are lives of leases not having their lives assured. In all cases of neglect of this kind, the advanced rent which might otherwise have been provided for becomes a burden on those who, it may be, were long used to small exertion, but who, if they had a little exercise in additional yearly outlays, might have been better prepared for the lapse of a lease of a cheap farm, to be followed by a high rent.

Besides, amongst tradespeople the neglect of life assurance, which is a truly Irish failing, is not only palpably manifested, but sorely felt in all parts of the country—even Ulster not excepted. Life assurance must either precede or accompany building societies, and other organisations by whose operations banking has been so much aided in Great Britain, and to meet the necessities of which so many banks have been aided in the earlier stages of their progress. We consider the idea so little known, except in Ireland, that the duty to insure one's life is only imperative when one grows sickly, ought to be combated in all possible ways, if for no other reason than the evils it brings on the families of those who are its victims, and the loss and litigation which follow attempts made to embody it in action.

There is, probably, no subject so generally known elsewhere, and yet so little understood, except amongst a few in this country, as the nature and objects of life assurance. Certainly, there is no topic by whose propagation so much good might be promoted which is so little heeded, and about which so few concern themselves. It is, however, if possible, even worse in respect to fire insurance. Many persons suppose they really gain all they can get by the greatest exactions possible in cases of a policy becoming a claim. This fact speaks to others, and leads to the conclusion that the public mind greatly needs instruction on this subject. The losses borne by assurance companies fall indirectly on the assurers, which is at once rational, fair, just, and what ought to be seen into, but which seems to be wholly overlooked. The rate per cent. on fire risks is regulated less by probable losses than by the actual experience of the several companies. It is also a sad and disagreeable fact that risks in all other respects alike are from 20 to 30 per cent. higher in Ireland than in England. This difference is all the more to be deplored because it is even less than ought to be laid on, were Irish business isolated altogether from British.

Practically, over and above these relations to banking, both life and fire assurance ought to be taken up and carried on more fully by Irish companies, whose shareholders, in the nature of things, would live amongst the people on whose lives proposals would be made, or on whose property risks would be taken.

The tens of thousands sent out of Ireland every year for life and fire premiums, if spent at home, would have to find investments which would in turn eventuate in many improvements like those to the effecting of which this money is now applied elsewhere.

As a class, the directors and managers of assurance business in Ireland need stirring up as much as any other, but, as in all other classes, there are creditable and marked exceptions.

No one acquainted, however, with Irish cases of both life and fire business, can fail to notice that, for want of really practical men in the several departments, both branches of this most important business is being badly done, to an extent which removes it beyond the range of the several companies' own concern to that of the whole community. On the management in accepting, refusing, fixing rates, and settling claims, depend both the rates of premium and the amounts to be divided periodically ; therefore offices should be filled by merit, none by favour—but all persons fit to do their duties, rather than by those who neither desire to know their business well, nor do what they know properly. As assurance companies may be the means in this country, as they have been in others, of much good, and as the chief reason why they do not progress is the want of truly efficient men in the several more important departments, it became a public duty to notice this defect in management, in the hope a remedy may be attempted, and reform achieved. The best remedy would be a few new Irish companies.

CHAPTER XXIV.

CONCLUSION.

To impose order in the development of Ireland's industrial resources would unquestionably enhance profits, but would also as certainly interfere with the rapidity of progress. Therefore, the lesser of the two evils must be chosen, which means that no one should wait for another, but every man, in whatever rank, do in the way of industrial development whatever he can, and do so at once.

Shopkeepers say it would be a breach of good order if their class bestirred themselves before merchants had shown them the example, and these in their turn say the landowners and gentry ought first to move, whilst even that class roll over the onus of the first efforts upon some other class in this country, or on the English and Scotch.

No man ought to hesitate in the discharge of those duties which rest upon himself, because others, on whom like duties may press even more heavily, fail to feel their weight, or to meet their claims. Yet, unhappily, such hesitation is to be seen in every part of the country, and amongst all classes ; and, worse than all, it is sometimes considered an evidence of superior knowledge, and is often sanctioned by those reputed wise. Nevertheless, it is really a matter of as much moment, so far as it affects the case, whether some one on the coast of the county of Galway begins to catch and cure fish according to the opportunities presented by the abundance of supply, or that some enterprising capitalist takes mill-sites and erects factories on the Corrib. In either case industrial development is promoted—the only difference is the extent of influence the respective operations are likely to have on the general interests of the country. Nor would it make any difference to the argument were the coast of Donegal, or that of any other littoral county in Ireland, substituted for Galway ; or that the Lee, the Shannon, or the Liffey, were named instead of the Corrib.

It seems strange, in the face of several successful efforts lately made in the least likely places, that anyone should continue to oppose feasible projects on no better grounds than the supposed propriety of putting off one thing in favour of another, lest the people appealed to for aid might be alarmed. Yet such flimsy excuses have from time to time prevailed against the most hopeful undertakings. The effort to get up a cotton manufacturing company in Galway being discouraged in favour of the packet station which is even yet hardly safe, is a case in point.

The year 1862 is unquestionably a crisis in Irish affairs. This one year will affect the industrial interests of the future more than they have been affected in the whole of the past thirty years. The cotton crisis in England has opened a field for the fair trial of flax, and also the introduction of cotton in Ireland, which, if improved, may raise this country to the highest position in the United Kingdom. But the opportunity implies also a danger that if it be allowed to slip it may never come again. Ireland's next opportunity will be when the fuel question will drive the factory owners to water-power, as scarcity of cotton has driven them to all possible sources of supply, and to every species of substitute. The present opportunity in comparison with that future, is not only better because it is now available, but also because the present being improved the future will be anticipated. Besides, that which is to come may be long postponed, and before it comes it may be rendered comparatively valueless by the rapid progress which may be made in other places whose water-powers are equal to those of Ireland.

Farmers, above all, have a great work to do this year. It would be quite possible to make the yield of their holdings, be they large or small, £2 or £3 per acre more than they have had on the average these last ten years. Agriculturists ought to decide now upon sowing flax, for example, to the greatest extent their soils will permit. The extent of the flax cultivation ought also to be published definitely by each organ of the press in every county, making the intentions of farmers within the bounds of their circulation known, so that persons interested as speculators in flax-seed or as flax-spinners might have sound data. On the average, the soils of Connaught are better than those of Munster, and these are superior to Leinster and even Ulster. So far as steeping, bleaching, &c., are concerned, these may be done in any part of the country. Running water is not only suitable, but best for flax-steeping. The best foreign in the British market is steeped in a running stream. Many a ton is steeped in the Bann. But even were it not so, patent steeping, which is not too dear, may be safely and easily applied. But, although farmers may not signify their intention to grow flax so extensively as would induce northern flax-spinners to contemplate extension westwards, yet there is no justification in such a case for the merchants and capitalists in the west, to whom this trade is open, for putting off longer. In like manner, those who know the richness of the sea in fertilising agents for the land are utterly inexcusable if they postpone the utilisation of sea-plants and inedible fish for manure. Neither would there be the shadow of excuse for negligence in those on whom rests the onus of turning peat-bogs to use, or minerals, or clays.

In the history of past experience in railway undertakings, banking, flax-spinning, ship-building, coal-mining, fisheries, &c., as well as in the nature and value of resources still neglected, there is every encouragement for joint-stock companies coming to the aid of every species of enterprise, and in being applied in every class of case. The fact of these resources being spread over all parts of the country, and yet being available at certain most convenient places, and in quantities at each such place sufficient for profitable utilisation, may safely be used as a basis for those calculations demanded by commerce as a condition of risking money in untried undertakings.

It must not be supposed that Irish people do not desire, as much as English or Scotch people, to make money. The chief difference between the British and the Irish, in regard to money-making, is, that the latter rush too often, and even blindfold, to far-off places in quest of materials far less likely to yield good wages than those they have left at home. But the British usually work up all they have at home before leaving for any country, however attractive. The British people study most of all how to bring the materials of other lands to aid the development of their home resources, and least of all how to make themselves possessed of such resources, if they are to be had in places far off. Hence Britain, South and North, has become the market for labour, skilled and unskilled, and the source of manufactures of many sorts, for the production of which the country possessed but few materials.

The resources of manufacturing and commercial industry in Ireland under such treatment for the next ten years as those of England and Scotland have had for the past seventy, and especially the past thirty, would yield as much permanent good to all concerned in their development as has been yielded in the whole of the greater period in the other cases.

The call to action suggested by a consideration of Ireland's state and prospects, though addressed primarily to the proprietors of lands, and the moneyed class of the country, because they would benefit first and most, has also a secondary and by no means an unimportant aspect for the manufacturers and merchants of Great Britain. By industrial development the Irish gentry would gain by the enhancement in value of their property, whilst the classes named, in England, would gain even more by the saving in the cost of producing those goods which form so large a portion of their trade. It would be no drawback on the value of such development that its introduction, like that of railways, would entail an apparent loss. Ruralising the textile trades of the United Kingdom would, of itself, be a sufficient equivalent for the transfer of cotton machinery to Irish water-power mills. We are not aware of the value of that machinery in each mill, nor of the number of mills, nor yet of the enormous sum the taking it down would cost, and the injury it would suffer by removal, as well as the expense of transit, and yet, although these would add up to 10 to 15 per cent. on £20,000,000 or £30,000,000, yet so much larger is the amount that might be fairly placed over against this one inevitable result of Irish industrial development, that the sacrifice would be small as compared with the gain resulting from its being made. How many of those now elevated to high positions by the gains of cotton-spinning within the last 60 years, and how many of the rich men now residing in England who made their money by manufacturing, had, when they first invested their capital in the business, that special knowledge of the trade which, it is assumed, ought to be had in Ireland before any movement is made, is a question the answer to which might astound many, because this answer would be not more than 2 in 10 on the average of the time named. To ask how many of those who have come to Ireland from England or Scotland, and have regretted their change for reasons arising solely from their being in Ireland, might also elicit an answer that would startle some people on the other side of the Channel, as probably those would be so few as to be scarcely worth mentioning at all. But Ireland has got so far in the right direction that few, if any, expect English or Scotch to settle in the country unless they see their way to do so on sound principles of commercial policy. Therefore, any one who now comes may be looked upon as adding his testimony to the fact that Ireland is a rich field for industry.

We now bring these chapters to a close. It is our sincere hope that, amid all their imperfections, they stir up inquiry among many of the classes of persons whom they are intended to benefit. Their large circulation, both in the columns of the "Evening Mail," and of the journals into which they have been either wholly or in part copied, has, doubtless, given the writer's views much influence where they were sound and suggestive. The making of each chapter a newspaper article rendered it impossible to do more than touch upon a number of points upon which a volume almost might with advantage be written. But the author has had the satisfaction of knowing that his meaning has very seldom been misconceived, and that his object has never been misinterpreted. Since he began to write, the most influential journals in the country have taken up several of the topics to which, in the accident of time, he was the first to direct attention. The *Times*, for example, has published several extremely able and timely articles on the flax question, every argument of which, mainly addressed to England, is still more applicable to this country. In like manner the same journal has spoken fitly upon the fuel question, throwing out hints of the revolution that is approaching in that department of industry, which Irishmen would do well to ponder.

Finally, the author hopes that his exertions may not prove in vain. Indeed, to a large extent he knows they have already been useful in proving their efficiency. There is a steadier and more practical attention paid to the subject now than ever before, and though the modes we commend of bringing capital to Ireland, or calling forth the capital already in the country, for large local enterprises, must in the nature of things be slow, the first step to that end must certainly be with public discussion as the " Evening Mail" has encouraged by the chapters now brought to a close.

APPENDIX.

———◆———

FLAX CULTURE—ENGLAND COMPARED WITH IRELAND.

To make room for an increase of 850,000 acres of flax in Ireland would displace oats to the extent of 500,000, and potatoes to the extent of 350,000 acres. But as the potato is a profitable crop, it might be as well to assume the supplanting of wheat, rather than potatoes. But, for a more important reason—namely, to bring the comparison as between England and Ireland to a position as nearly equal as possible—let it be supposed that to extend flax culture in either country would displace wheat, oats, beans, and peas only, and not potatoes, turnips, grass, or other crops ; also that this displacement would be according to the average proportion in which such crops are cultivated in the respective countries. For every acre of flax in England, therefore, 6-10ths of wheat, 2-10ths of beans and peas, and 2-10ths oats would be displaced, and in Ireland, on the same principle, 5-10ths oats, 4-10ths wheat, and 1-10th beans. Taking the average nett profits to be on these crops respectively, after paying every expense, on oats, £3 ; on wheat, £4 ; on beans and peas, £2—and that these averages apply alike to both countries, the displacement in England in favour of flax would be, therefore, of crops yielding £3 on the average, and in Ireland of crops yielding £2 12s. per acre per annum. But Ireland, in present circumstances, cannot grow either as good crops, nor get as high prices for wheat as England. Nor would the change of Ireland to a manufacturing country, although it would imply the consumption at home of her crops, and the greater import of foreign grain, alter the respective rates for wheat in England and Ireland. Irish farmers may not safely lose sight of the fact, that wheat can be imported into Ireland 2s. to 3s. per quarter, or say 10s. to 15s. per ton, cheaper than in England ; whereas Irish farmers may make from £1 to £1 10s. per acre more of flax crops than it would be possible for English farmers to make, except (as is forcibly suggested by the respective climates), flax were cultivated in Ireland for the fibre only (except the use of bolls for feeding), and in England chiefly, even at the expense of the quality of the fibre, for seed.

Whatever may be the effect of the cotton crisis on the price of flax for the next five years, English agriculturists cannot afford to lay aside wheat, beans, and peas crops, to an extent above half that to which Irish agriculture ought to make oats, wheat, and beans, secondary crops. Ireland is just as naturally a great flax field, to the extent of one-fifth or one-sixth of her cultivated area, as are the lands in certain climates suited for cotton. But, taking the respective averages as they are, and comparing them with that which flax may yield, there is a difference in favour of flax over any of the crops named enough to induce both English and Irish farmers to grow it to the fullest possible extent.

Taking a statute acre of land fit to grow flax at £2 10s. rent; seed, £2; horse and manual labour in preparing the ground and sowing the seed, £2; weeding, pulling, and drawing to the water, or rotting artificially, £2; taking out of water, bleaching, and carrying to the mill, £2 10s.; beetling, scutching, &c. (depending on the yield), say £2 10s. Suppose the yield to be 5 cwt. to 6 cwt., mean, 5 cwt. 2 qrs., and the price to be 70s. to 80s. per cwt., the amount, plus £3 for bolls, and minus £1 for artificial manure, to leave the land in as good condition as it would have been after a crop of oats or wheat, would be £20 12s 6d.

But, supposing nothing made of the bolls, the yield would be £18 2s. 6d., and the cost £13 10s.—being at the lowest £4 12s. 6d., or, with the bolls, £7 12s. 6d., against £3 in England, and £2 12s. in Ireland, per acre, average profit from other crops.

In making this calculation nothing is allowed for scutching-tow, which may be taken at £1 per acre, making the profits of course so much more.

Machinery in several of the branches of manipulation might yet be adopted so as to reduce the outlay below £13 10s., probably to £10 per acre. Besides, be the price what it may afterwards, flax is likely to keep its average of £70 to £80 for the next five years, and, if yarns be woven by power-looms, may rise up to £100 for such really fine, strong fibre, as Connaught could, of all places in the United Kingdom, the best produce. The economy of the shoves for manure seems impracticable, except by charring them, which might be done in a close fire or oven beside each mill. Any one who has observed the fertilizing powers of the ashes of shoves must, by such a fact, be convinced that, to put the shoves through such a process would pay well, as, in any case, they are worthless except for fuel. and for that worth very little. This is one feasible way to bring the valuable mineral powers of this flax waste to bear upon the restoration of the soil. Another and more simple plan would be to use the shoves for bedding pigs. Each acre of flax would yield in long shoves (like tangled oaten straw) about one ton, and short shoves (like sawdust, about half the quantity. Either of these used as a base of twelve to thirty inches thick for pigs' beds, would last under straw a whole winter, and would then make the best of manure. But the probable quantity of flax likely to be required is the next matter of moment.

According to calculations based upon a reduction of the prices of cotton to 1s. per lb. for New Orleans "fair" as a standard, the number of flax-spinning mills, with weaving factories attached, in the United Kingdom might be safely increased 500 to 600, which would increase the demand for flax up to the produce of 1,500,000 acres, of which probably foreigners may send us 1-5th, and the rest we must grow.

With these facts before us, there is really no excuse for delay in regard to flax culture, especially as we learn, with great pleasure, from the columns of the *Northern Whig*, that power-loom machinery is being brought into Belfast, and mills long standing are being put into working order again.

For the sake of readers generally we have taken the following directions for flax-growing from one of the annual reports of the late ROYAL SOCIETY FOR THE PROMOTION AND IMPROVEMENT OF THE GROWTH OF FLAX IN IRELAND, to whose labours we trace much of that preparedness to improve the present opportunity for the extension of flax-growing, so manifest in most parts of the west, and also in many places in the south of this country. We ought, perhaps, to apologize to our northern readers for taking up so much space with matters of detail with which they are so familiar; but we trust the common interest all parts of the country have in flax-culture will be a sufficient excuse, and those very few original remarks we have ventured amongst quotations from so high authority, will render the whole deserving of perusal from the most experienced flax-grower.

We notice also, with great pleasure, the frequent allusions to the subject of flax culture recently made by Irish journalists. We are happy to say the testimonies of these to the importance of this matter are almost unanimous; and so ably written, so well-timed, and so oft-repeated, as to be worthy of all praise for every reason which could be associated with the discussion of the following momentous questions:—How many acres of flax ought the Irish farmers to grow? How much more labour will flax require than oats, wheat, or other crops? How will flax pay in comparison with those crops its increased growth must displace?

DIRECTIONS FOR THE PROPER MANAGEMENT OF THE FLAX CROP,

Partly those compiled by the Committee of the late Royal Society for the Promotion and Improvement of the Growth of Flax in Ireland, and partly the result of our own Inquiries of Practical Farmers.

SOIL.—By attention and careful cultivation, good flax may be grown on various soils; but some are much better adapted for it than others. The best is a sound, dry, deep loam, with a clay subsoil. It

K

is very desirable that the land should be properly drained and sub-soiled; as, when it is saturated with either underground or surface water, good flax cannot be expected.

Without method, there cannot be success—different soils require a difference of rotation. Having often met with the statement that flax is a peculiarly exhausting crop, we have taken special pains in preparing matter for this work, to learn, from practical farmers, the result of their experience—which is, according to evidence obtained from both large and small farmers, that FLAX IS NOT *an* EXHAUSTING *crop in any special sense.*

PREPARATION OF THE SOIL.—One of the points of the greatest importance, in the culture of flax, is by thorough draining, and by careful and repeated cleansing of the land from weeds, to place it in the finest, deepest, and cleanest state. This will make room for the roots to penetrate, which they will often do to a depth equal to one-half the length of the stem above ground.

After wheat, one ploughing may be sufficient, on light friable loam, but two are better ; and, on stiff soils, three are advisable—one immediately after harvest, across the ridges, and two in spring, so as to be ready for sowing in the first or second week of April. Much will, of course, depend on the nature of the soil, and the knowledge and experience of the farmer. The land should be so drained and subsoiled, that it can be sown in flats, which will give more even and much better crops. Subsoiling should not be done at a less interval than two years prior to the flax crop. This gives the land time to consolidate. But, until the system of thorough-draining be general, it will be necessary, after oats, to plough early in autumn, to the depth of six or eight inches. Throw the land into ridges, that it may receive the frost and air ; and make surface drains to carry off the rains of winter. Plough again in spring, three or four inches deep, so as to preserve the winter surface for the roots of the flax. The spring ploughing should be given some time before sowing, to allow any seeds of weeds in the land to vegetate, and the harrowing in of the flaxseed will kill them, and save a great deal of after weeding. Following the last harrowing it is necessary to roll, to give an even surface and consolidate the land, breaking up this again with a short-toothed or seed-harrow, before sowing, which should be up and down, not across the ridges or anglewise.

SOWING.—The seeds best adapted for the generality of soils is Riga, although Dutch has been used, in many districts of country, for a series of years with perfect success. American seed does not generally suit well, as it is apt to produce a course, branchy stem. If used, it should be on deep, loamy soils. In buying seed, select it plump, shining, and heavy, and of the best brands, from a respectable merchant. Sift it clear of all the seeds of weeds, which will save a great deal of after trouble, when the crop is growing. This may be done by fanners, and through a wire sieve, twelve bars to the inch. Home-saved seed has produced such excellent crops, that it is strongly recommended that every farmer should only sow, each year, as much foreign seed as would produce a sufficient quantity for his flax crop of the following season. The produce of seed averages about twelve bushels the statute acre, so that the seed saved off one statute acre would sow about five. It must not be overlooked that so much depends on good seed, that in a case in which so much depends, inspectors competent to judge (as we have already said) ought to "*pass*" every bushel of seed allowed to find its way to the soil. Belfast merchants do not hesitate, at the expense of sending first-rate judges to Russia and Holland, to inspect what they have to pay for, and therefore it is but fair they should not be allowed to sell had seed at full price, which, under such circumstances, it may be fairly assumed, came to their possession at a very cheap rate. The thinner portion of the crop is the best for "saving seed" from, as, when flax grows thin, it produces much seed. This plan, besides the saving effected in the price of foreign sowing seed, would effectually secure the farmer from any danger of loss from fraudulently made-up seed. It will be best, in most cases, to use the seed which is saved from home grown seed in the following year, for feeding, or to sell it for the oil mills, although it often produces good crops. The proportion of seed may be stated at three-and-a-half imperial bushels to the Irish or plantation acre, and so on, in proportion to the Scotch or Cunningham, and the English or statute acre. It is better to sow too thick than too thin ; as, with thick sowing, the stem grows tall and straight, with only one or two seed capsules at the top ; and the fibre is found greatly superior, in fineness and length, to that produced from thin-sown flax, which grows coarse, and branches out, producing much seed, but a very inferior quality of fibre. The ground being pulverised and well cleaned, roll and sow. If it has been laid off without ridges, it should be marked off in divisions, eight or ten feet broad, in order to give an equable supply of seed. After sowing, cover it with a seed harrow, going twice over it—once up and down, and once across or anglewise ; as this makes it more equally spread, and avoids the small drills made by the teeth of the harrow. Finish with the roller, which will leave the seed covered about an inch—the proper depth. The ridges should be very little raised in the centre, when the ground is ready for the seed, otherwise the crop will not ripen evenly ; and, when land is properly drained, there should be no ridges. The sowing of clover and grass seeds along with the flax. is not advised, and when it can be conveniently avoided, as these plants always injure the root ends of the flax. But carrots may be sown, in suitable soils, in drills, so that the person pulling the flax may step over the rows, which may be afterwards hoed and cleaned, and should have some liquid manure. A stolen crop of rape or winter vetches, or of turnips of the stone or Norfolk globe varieties, may be taken, after the flax is pulled.

MANURE FOR THE FLAX CROP.—Recent chemical investigations have shown that the fibre of flax does abstract from the soil certain matters, although not in so large a proportion as several other commonly cultivated crops. To supply to the soil all the matters which the plant requires, with the exception of what is abstracted by the seed, which should be returned by saving the latter, and applying the manure of the cattle fed upon it, or an equivalent if the seed be sold, so as to leave the land in the same state of fertility as before, the following compound, proposed by Professor Hodges as a manure, which may be sown broadcast on the land, prior to the last harrowing before sowing the flaxseed—viz., per statute acre :—Muriate of potash, 30 lbs.; chloride of sodium (common salt), 28 lbs.; burned gypsum, powdered, 24 lbs.; bone dust, 54 lbs.; sulphate of magnesia (Epsom salts), 56 lbs. Since the eminent and learned Dr. Hodges gave the above recipe, the practicability of producing manure specially rich in the elements abovenamed has been rendered simple, especially by the operation of M'Ardle's patent on seaweeds. We feel it our duty, however, for the sake of such of our readers as do not know of Dr. Hodges' services to agriculture in Ireland, to say that he has done more for scientific farming than, probably, any living man, as is testified by statistics frequently quoted, labours incessant and painstaking, which have been rendered the more effective, because they have been divided between the duties of the professor's chair in the Queen's College and the interests of the Chemico-Agricultural Society of Ulster.

WEEDING.—If care has been paid to cleaning the seed and the soil, few weeds will appear; but if there be any, they must be carefully pulled. It is done in Belgium by women and children, who, with coarse cloths round their knees, creep along on all-fours. This injures the young plant less than walking over it (which if done, should be by persons whose shoes are not filled with nails). They should work, also, facing the wind, so that the plants laid flat by the pressure may be blown up again, or thus be assisted to regain their upright position. The tender plant, pressed one way, soon recovers; but if twisted or flattened by careless weeders, it seldom rises again.

PULLING.—The time when flax should be pulled, is a point of much nicety to determine. The fibre is in the best state before the seed is quite ripe. If pulled too soon, although the fibre is fine, the great waste in scutching and hackling renders it unprofitable; and if pulled too late, the additional weight does not compensate for the coarseness of the fibre. It may be stated that the best time for pulling is when the seeds are beginning to change from a green to a pale brown colour, and the stalk to become yellow, for about two-thirds of its height from the ground.

RIPPLING should be carried on at the same time, and in the same field, with the pulling. If the only advantage to be derived from rippling was the comparative ease with which rippled flax is handled, the practice ought always to be adopted; but, besides this, the seed is a most valuable part of the crop, being worth, if sold for the oilmill, £8 per acre, and if used for feeding stock of all kinds, at least £4 per acre. The apparatus is very simple. The ripple consists of a row of iron teeth screwed into a block of wood. The best ripples are made of half-inch square rods of iron, placed with the angles of iron next the ripplers, 3-16ths of an inch asunder at the bottom, half an inch at the top, and eighteen inches long, to allow a sufficient spring, and save much breaking of flax. The points should begin to taper three inches from the top.

WATERING.—This process requires the greatest care and attention. River water is the best. If spring water has to be used, let the pond be filled some weeks or months, if possible, before the flax is put in, that the sun and air may soften the water. That containing iron or other mineral substances should never be used. If river water can be had, it need not be let into the pond sooner than the day before the flax is to be steeped. The best size of a steep-pool is 12 to 18 feet broad, and 3¼ to 4 feet deep. Place the flax loosely in the pool, in one layer, somewhat sloped, and in regular rows, with the root end underneath ; the tie of each row of sheaves to reach the roots of the previous one ; cover with moss sods, or tough old lea sods, cut thin, laid perfectly close, the shear of each fitted to the other. Before putting on the sods, a layer of rushes or rag-weeds is recommended to be placed on the flax, especially in new ponds. As sods are not always at hand, a light covering of straw may do, with stones laid on it, so as to keep the flax just under the water ; and as the fermentation proceeds, additional weight should be laid on, to be removed as soon as the fermentation ceases, so as not to sink the flax too much in the pool. Thus covered, it never sinks to the bottom, nor is affected by air or light. A small stream of water allowed to run through a pool has been found to improve its colour. In this case, if the pools are in a line, the stream should be conducted along the one side, and run into each pool separately, and the water of each pool run off, along the opposite side, in a similar manner. It will be sufficiently steeped, in an average time, from eight to fourteen days, according to the heat of the weather and the nature of the water. Every grower should learn to know when the flax has had enough of the water, as a few hours too much may injure it. It is, however, much more frequently *under*-watered than *over*-watered. The best test is the following :—Try some stalks of average thickness, by breaking the *shove*, or woody part, in two places, about six or eight inches apart, at the middle of the stalk ; catch the broken bit of wood, and if it *will pull freely out, downwards, for that length, without breaking or tearing the fibre, and with none of the fibre adhering to it*, it is ready to take out. Make this trial every six hours, after fermentation subsides, for sometimes the change is rapid. Never lift the flax roughly from the pool with forks or gripes, but have it carefully handed out of the flax drain by men standing in the water. It is advantageous to let the flax drain twelve to twenty-four hours, after being taken from the pool, by placing the bundles on their root ends, close together, or on the flax, with the slope ; but the heaps should not be too large, otherwise the flax will be injured by heating.

There are chemical systems of steeping or retting flax, on a large scale, now in operation, of the merits of none of which do we venture to treat, being convinced that these, like all similar inventions, will be the more fairly tested the more generally they are known ; but at the same time take leave to suggest to the several patentees, proprietors, and may we not hope inventors in embryo, (now that it seems as if in this year of 1862, 400,000 to 500,000 acres are likely to be sown, and more and more yearly till the right quantity is reached,) that April, May, and June at the latest, are the months in which they ought to begin to exhibit, advertise, or otherwise make known their proposed plans for the substitution of the old system, because farmers ought to have time to consider and take counsel on a matter so momentous to their own and their country's interest before adopting any new plan, however well recommended or likely to answer.

Spreading.—Select, when possible, clean, short, thick pasture ground for this operation ; and mow down and remove any weeds that rise above the surface of the sward. Lay the flax evenly on the grass, and spread thin and very equally. If the directions under the head of rippling have been attended to, the handfuls will come readily asunder, without entangling. Turn it two or three times while on the grass (with a rod about eight feet in length, and an inch and a half in diameter), that it may not become of different shades, by the unequal action of the sun, which is often the case, through inattention to this point. Turn it when there is a prospect of rain, that the flax may be beaten down a little, and thus prevented from being blown away.

Lifting.—Six to eight days if the weather be showery, or ten to twelve if it be dry, should be sufficient on the grass. A good test of its being ready to lift is to rub a few stalks from the top to the bottom, and when the wood breaks easily and separates from the fibre, leaving it sound, it has had enough of the grass. Also, when a large proportion of the stalks are perceived to form a *bow and string* from the fibre constructing and separating from the woody stalk. But the most certain way is to prove a small quantity with the hand-break or in a flax mill. In lifting, keep the lengths straight and the ends even, otherwise great loss will occur in the rolling and scutching. Let it be set up to dry for a few hours, and afterwards tie it up in small bundles ; and, if not taken soon to be scutched, it will be much improved by being put up in small stacks loosely built, with stones or brambles in the bottom to keep it dry, and allow a free circulation of air. Stacks built on pillars would be the best.

Drying by fire *is always most pernicious.* If properly steeped and grassed, no such drying is necessary; but to make it ready for breaking and scutching, exposure to the sun is sufficient. In some districts it is put to dry *on kilns*, in a damp state, and is absolutely burned before it is dry, and the rich oily appearance of the flax is always greatly impaired. On this point it is scarcely possible to speak too strongly, as the flax is either destroyed or rendered not worth one-half of what it would be if properly prepared.

BREAKING AND SCUTCHING.—If done by hand, should be on the Belgian system, which is less wasteful than that practised in Ireland. If by milling, the farmer will do well to select those mills in which the improved machinery has been introduced.

SEAWEED.

BUT few questions are of more moment than what is the probable quantity of seaweed which may be calculated upon on the Irish coast. It would be impossible to answer this question accurately, but it would be alike impossible by any means within the control of human power to exhaust the supply cast by every tide upon the Irish shores.

Taking the measurement of a line describing the shore and the banks around all the interstices from which the weed is washed on the land, from Valentia (or Dingle Bay) to Carrickfergus, it would measure 1,500 statute miles, on each mile of which no less a quantity, on an average, than 3,500 tons are cast yearly, or 5,250,000 tons in all, to which may be added two to three millions cut weed (mean 2,500,000) or 7,750,000 tons of weed, equal in iodine, bromine, and the more valuable salts to the best weeds of Scotland, but superior to that cast on about 500 miles of coast line from Belfast Lough to Valentia, and which yields on the average about 2,000 tons per mile, or 1,000,000 tons. In the same line might also be had 500,000 tons of cut weed—making in all 9,250,000 tons as the yield in Ireland annually.

The quality of the weed on the east and south-east coast of Ireland is as rich in the more valuable salts as weed is on the average in any part of England. Both English and south-east Irish weeds are quite equal to those of France and Belgium, and very superior to those of Spain. For fertilising purposes, so long as weed is used in a wet state, it makes very little difference between one kind and another, but when it is subjected to chemical treatment a greater or a lesser quantity, according to the amount present of the ammoniacal salts, may be left in manure; so that, under scientific treatment, the difference is as much in favour of the west and north-west of Ireland weeds for manure-making as for any other use.

The above estimate would allow 1½ tons to every acre of cultivated lands in Ireland. But it may even be said that, if required for the purpose, there is as much weed to be had in Ireland as would give the manurial products of two tons to every acre of land under culture. This, in its relation to farming, is a fact that ought not to be overlooked. The supply in this case coming naturally may not be controlled artificially, and is, therefore, independent of either commercial or political changes; and although it has its fluctuations, yet it always obeys the laws of average. These calculations, although unsupported by any demonstrable data, may be tested by anyone who will measure the coast line and observe the quantity left by each tide. Like the estimates as to the time coal is likely to last out, those as to weeds are open to considerable difference of opinion; but, unlike coal calculations, they may be reduced to an approximation near enough for all practical purposes. The tendency of all this evidence is towards conclusions calculated to give unbounded confidence in the outlay of money to any extent in works dependent for their raw material on the supply of seaweed, and also to warn persons possessed of royalties not to demand high rates, as the supply is so large and the distribution so general—those who require to buy rather than those who have weed to sell can fix the price. It is to be hoped attention may be given to these facts and their importance duly improved by capitalists, chemists, and landed proprietors in this country.

FISHERIES.

THE most useful mode of dealing with the Irish Fisheries question is to dismiss it summarily, by saying that the supply of every kind of edible and inedible fish is abundant, the fishing grounds are easy of access, and the demand for the greatest quantity of fish which could ever be taken, even under these peculiarly advantageous circumstances, would be found equal to the supply. This places the subject in its right position. Still more—each one of the foregoing statements being supported by evidence, leaves no excuse for those who hesitate to invest money in self-supporting organisations—whether co-partnerships or joint-stock companies, with limited or unlimited liabilities, it matters not, for fishing operations.

The chief cause, if not the only one, whose operation has kept the Irish fisheries in the condition they have been in so long, is the expectation of Government aid, or at least some assistance from rich people, in the name of encouragement to Irish enterprise or charity to the poor fishermen of the west.

So soon (since ever the question of the development of fisheries became a leading topic) as a paper on fisheries was read at some meeting, or any other praiseworthy exertion made to propagate the truth in reference to this subject, or to expose the anomalies of Irish merchants importing Scotch fish, leading articles, ably written, followed up the movement, but these, with few exceptions, amounted to no more than a whine or a grumble, and seldom failed to counteract the good done by the efforts to the approval of which they seemed to be directed. Sometimes other evils followed as a consequence of ill-timed and badly-judged advocacy of those very expectations, which, had the press been in a healthy state on this subject, would have been denounced as the only source of that apathy under which Irish fisheries were kept in an undeveloped state. How could it be expected that shopkeepers, traders, farmers, or others who had sums from £20 to £200 to invest, would follow the example of their respective classes in Scotland, England and Wales, by buying boats, tackle, and all appliances for fishing the Irish waters, whilst agitations were constantly at work, whose promoters promised to get money (which they knew would never be given) from Government, to do wonders which only existed in their impractical imaginations, and were only talked of, or written about for personal or party purposes? Hopes of aid, aroused only to be broken on the one hand, and fears which menaced but could never hurt enterprise on the other, were no less united in their baneful effects on the spirit of industry in this department than they were in the common origin as the fruits of a gross mistake as to how fisheries ought, or were ever likely, to be developed in Ireland. Large capitalists were also kept out of this trade, inviting as it was, for reasons if not exactly the same, yet identical in their nature, and equally evil in their effects. To disabuse the public mind from the false notion that there is any necessity for special aid, and also from still greater error that either precedent in the past history of this country, proceedings now followed with good results in other Governments, or even the social condition of the "poor" ought to divert the minds of moneyed people in Ireland from the simple truth that on ourselves alone rests the responsibility of developing our long-neglected fisheries, is the only work now left for either the Editor, the Author, or the Orator, and, this being done, and done faithfully, plainly, and persistently, the time cannot be far distant when Irish capitalists and others, like those of England and Scotland, will make fisheries a branch of industry, and a source of employment for talent, money, and labour, for the sake of reward, and for that reason to the exclusion of all others.

IRISH FREEHOLD LAND COMPANY.

THE advent of an Irish Freehold Land Company under the Limited Liabilities Act, whose business would be to buy estates and re-sell, let on long leases or in perpetuity on equitable terms to occupying tenants or others, would be specially opportune at the present time. Such a company would find in the lands offered for sale by the Landed Estates Court abundance of material for operation, and might be so conducted as to yield large profits to shareholders. The conditions suggested for joint-stock companies generally (page 80) should be rigidly enforced in a land company, and which, being brought to bear, would, undoubtedly, result in the establishment of a new and most profitable line of business in this country.

Probably the only barrier to the introduction, and the chief danger to the prosperity of such a company, are in the unsound state of public opinion on the " Land Question " in Ireland.

Judging of what appears in print on this topic, in the light of facts not only visible on all sides, but too palpable to be overlooked, we are constrained to say that not only is the introduction of such a necessary addendum to the Landed Estates Court (as a large commercial company for trading in lands) postponed from year to year by the propagation of erroneous views on the relation of land to capital, and of both to labour, but to the same source must also be traced very many evils under which both the social and commercial interests of the country are suffering, and we regret to say most severely at present. We see it stated, if not in so many words, yet in effect, that land is raised to artificial rents, and held at such prices by means resorted to unfairly by " landlord interest " to the injury of the occupiers in, and the tillers of soil. But even those who seek proofs of this or kindred assumptions cannot find them in any part of the country. The fact is, occupiers have driven up the " good will " of lands, and most of all, where tenant-right prevails, to a fee-simple price, in which evil, prevalent as it is in one entire province, and its reflexive effects in the other three, may be found the common centre around which revolve most of the strictly agrarian grievances of Ireland.

It is asserted, moreover, that for the want of greater security of tenure, the condition of the farming interests is in a state of hopeless decline.

But the truth is, however, that for want of that comparative independence, which is never to be found except where there are manufactures, rather than to insecurity of tenure, is that indolence in the tillage of lands traceable, which we admit is a real Irish grievance.

A far greater cause of complaint arises from the fact that persons unused to look at matters minutely pay for a few acres of land (or the "good-will," which is a right to be "a tenant-at-will") prices from £30 to £45 per Irish acre, a part of which money is often borrowed at 6 per cent.;—to pay the interest on which, and a rent, the capital being all sunk, even when the liberty to sell is allowed, must lead to loss, perhaps to ruin. The consequences of this evil, like those of many others, are attributed to insecurity of tenure, whereas their origin is in the abuse of the system of tenant-right, which gives a false confidence in the outlay of money, in buying what is supposed to be land, but is at the best only permanent improvements on land, which in any case may be taken possession of, without the violation of any legal right, by the landlord. It is not to the tenant-right custom, but to manufactures, the greater prosperity of Ulster is owing.

For example, let all that which manufacturing industry has done for that province be taken from the sum total of its comparatively prosperous condition, and there will be little left for the credit of "tenant right" worth notice. The fashion is, however, to give all to tenant-right, and nothing to manufactures; and though this is done for purposes apparently benevolent, yet doing so is so grossly against both truth and evidence as to be inexcusable even in those least acquainted with the real state of the case. Were the custom of tenant-right left to rest on the narrow basis which its own results have reared up for its support, it would soon be amongst the things that have been. Tenant-right is a most complicated and difficult system, and seems destined to remain merely a custom in some parts only of the north of this country, but is also not unlikely to be superseded by a healthier system so soon as manufactures give occasion to prove that "*labour* not *land*" gives liberty.

But the days have gone by for complicated arrangements, and therefore those now to be entered into must be simple. In like manner, also, the time is past for legislation in matters which, to be rightly managed, must be allowed to stand on foundations purely commercial. It is still more imperative that all efforts should be abandoned by which it is expected to cause landlords, in one part of the country, to follow modes of treating their tenants commended, mostly because others in different parts having adopted and found them feasible, or it may even be satisfactory.

Still more absurd is it, also, to conjure up imaginary *El Doradoes*, and then make rules by whose application they might, if people were passive, be brought into a temporary existence. The fact is, that in opposition to, and in despite of all that has been done to the contrary, landlords will be guided by their own ideas in letting property; and the only protection tenants can get from the possibility of their doing wrong is, in being placed in a position to be "the other part" in the bargain for land, whether by the year, or on leases for ten, thirty, or more years. English manufacturing progress gave farmers in that country the opportunity of choosing between land and manufactures, and the result has been, that landlord and tenant, across the channel, though not without their misunderstandings and mutual recriminations even in that country, let and take lands on much more mercantile principles than are customary in Ireland. But their position is no model, nor could even the benefits of their modes of action be taken advantage of in this country, except in so far forth as both countries are in exactly similar circumstances in all respects. But Ireland has it in her power now to do far more towards the bringing about a more perfect settlement of the land question in all its bearings, than England ever did, or had it in her power to do. It is quite practicable, in every sense, and altogether safe on commercial principles, to inaugurate a system of dealing in lands by which those who have money to invest might do so; and yet be able to re-sell their shares, so as to have their money at any time in case of need, as also for those who wish to make the most of their special knowledge as farmers, to get the advantages of their industry by buying lands, and paying for them by instalments, spreading over ten, fifteen, twenty, or twenty-five years; or by paying half purchase-price in such or any other manner, or by getting a lease in perpetuity, or a long lease at ordinary or existing rents.

To bring about all this by means of a company is feasible. To overcome all the difficulties in the way of doing it would be very easy. The probable results of such changes as would follow the earliest operations of such an undertaking would tell evidently in favour of the poorest holders of land in the country, and would tend directly to the elevation of the richer and more industrious to positions of independence, by putting within the reach of occupiers the means of being wholly or in part proprietors of the soil they tilled.

This Freehold Land Company project has not been hastily arrived at; on the contrary, it is the result of many years' careful consideration; and now that, before committing to these pages, it has been re-considered, we hesitate not to say, the only practicable and, in fact, the only necessary duty now devolving on the press, or any and all other exponents of principle for the guidance of public thought on the land question, advocate the adoption of commercial, and discourage all desires for parliamentary interference, so oft sought after for the settlement of unsettled points, as between the owners and the occupiers of land in Ireland. We cannot, in fidelity to the cause of truth, help mentioning, however, that to the absence of manufactures, an occasion has been given for most of the difficulties surrounding the land questions of this country. These have become more perplexing and disagreeable, because of the most remarkable anomalies in this country is, that, not unfrequently, persons of the class by whose monopolising greed and narrow-minded conduct manufactures are kept out of the country, are loud in pressing on those in power to interfere in favour of the tenant farmer, whom they pretend to commiserate, and yet whose only cause of complaint, over which he has not the power of controul, is in evils of which those his would-be benefactors are the authors—namely, the want of that circulation · of money which accompanies manufactures, and without which farmers can never do well, and with which we seldom notice them do badly. Surely the withering words of the poet are even too lenient, as applied to those who might expand manufactures, and through them enrich the farming interests, and will not, for mere selfish reasons, and yet cease not to disturb persons otherwise disposed to be content, by making them feel miserable and, consequently, careless. These are they who do, indeed,

"Compound for sins they are inclined to,
By damning those they have no mind to."

M'ARDLE'S PATENT.

THE history of the discovery, for the protection of which the discoverer (Mr. James Smith M'Ardle) has obtained letters patent, is like that of all great discoveries, not only little talked of, but even at the earliest stages of its existence less appreciated than matters of inferior importance.

Being fully acquainted with the various stages in the progress of this momentous matter since (in 1860) the patentee took the first steps in preparing his plans for active operation, we have been so often reminded, by what passed under our observation, of the early struggles (now matters of history) of Watt, Stephenson, Wedgewood, and others, to the operations of whose genius, under no ordinary difficulties, trade and commerce are so deeply indebted, that we believe it is only those who have studied these on the one hand, or such eminent chemists on the other, as those who have already borne testimony to the value of M'Ardle's invention, that can at all appraise its worth, or as yet assign reasons for working it fully out. But in other respects, also, M'Ardle's invention has many striking resemblances to those in regard to which we who now live and enjoy such benefits, wonder why those who lived at the time of their advent did not avail themselves more fully of the advantages they offered.

The application of steam to navigation, locomotion, and many other works in which it was once thought impossible to adopt it, is simply the more scientific employment of the common and abundant elements of coal and water; and yet the advantages to trade and commerce, and through them to civilisation and progress, arising from the use of steam, are beyond all possible calculations.

In like manner, the patented process now before us aims at no more than a scientific utilisation of sea-weeds, which, like water, is daily presented to our view, but, unlike coal, (the supply of which may be effected artificially,) comes naturally, and fails not twice daily to deposit its rich treasures on our coasts, in our bays and creeks, and may be had in many parts of this country growing in luxurious abundance on the rocks, in tide marks and shallow waters. Besides, doubtless, as persons who used to limit the utility of coal to the narrow sphere of domestic uses were the first to suggest the idea of danger in widening the range of demand; so are those now the first to be alarmed who suppose sea-weed ought to be left to the farmers who may be fortunate to live near the sea side, or to a few kelp-burners who visit our coast once a-year. These, for want of knowledge of facts, suppose the demand likely to arise under M'Ardle's process calculated to injure the interests of farmers and kelp-burners; whereas the very contrary will be the result.

In the supply of house coal, which, had it not been effected by the increased demand to meet steam purposes, would probably have fixed us often in a famine the last thirty years. There is a pertinent reply to the objections raised on the sea-weed question, which of itself ought to be sufficient, but which is irresistibly unanswerable, because seaweed subjected to this process becomes more productive of manure, and yields larger quantities of salts, to extract which the kelp is made; whilst of one (iodine) the quantity is 500 per cent. more, and another (bromine), which was not got at all, quantities large enough to repay nearly the whole cost of the raw weed are to be had, the qualities of all are necessarily improved, and the cost of the whole process GREATLY LESSENED.

In many other particulars might this invention, now only requiring money to work it effectively and profitably, and those others by the working of which thousands have been enriched, be compared, but the space at our disposal will not admit of saying any more than that M'Ardle's, like all the rest, was well-timed. The circumstances of the respective countries also called each forth, although the people (and we now state facts well known to all who know either case) did not for some time see the benefits such improvements offered, but rather opposed their introduction. We, therefore, take courage in commending M'Ardle's process to public support, because we regard the objections raised against it rather in its favour than otherwise. In doing so, we rely on inference from facts of a like nature, which any one may easily induct from the history of British commerce in the present century, whose leadings justify us in believing that the time will soon come when the most rigid economist will be convinced, by evidence he cannot question, that no such profitable speculation has ever been offered in Ireland as the working out the M'Ardle process in making iodine, bromine, manure, &c., from sea-weed.

WHILST the foregoing pages were passing through the press, a variety of circumstances, daily coming under our notice, have urged us to devote a portion of the appendix to a subject we would have made one of the most prominent topics in the book had those features of the case then existed which now demand our attention. But we cannot preface our remarks on this matter so well as by the following quotation from the columns of the " Dublin Evening Mail " :—

" We now say, in a sense different and better than that in which it has been said by foolish people before, "England's difficulty is Ireland's opportunity." For years past enterprising and patriotic Irishmen have deplored the neglect of our great industrial resources. Splendid streams, that would turn mills for an empire, are flowing idly and uselessly away. But what was to be done ? Cotton was king ; cotton-working was concentrated in Lancashire ; and though a few spirited manufacturers started successful mills in Ireland, there was still a certain disadvantage in isolation. But we have now the opportunity thrust upon us. Every day the cry in England grows louder for flax or other material available for textile manufactures. Ireland will always be an agricultural and pastoral country, and it is better that it should be so, for a mixture of the farming and factory interests is better for both. Flax-spinning has this great advantage, therefore, that it gives the farmer and the peasantry profitable work, while it sets mills going, which again employ the people, and increase the wealth and general prosperity of the country. We have often complained of the decay of our commercial and manufacturing interests, and this, too, not without reason. We point to our customhouses unemployed, to our noble harbours, whose waters are seldom disturbed except by coal and provision ships, and to our immense water-power left idle. The best thing we can do is now, when we have got the chance, prove that our complaints were sincere by a vigorous effort to set right what was wrong. If it be observed that Irishmen complain much of the poverty of their country, and yet neglect such an opportunity as at last presents itself, our enemies will certainly say that our grievances are due more to ourselves than to our situation."

The considerations which called these remarks (in August, 1862) have been greatly intensified, and now (January, 1863) require to be still more plainly dealt with. It is now more apparent than at that time that cotton is not king. Concentration of cotton manufactures in Lancashire is mainly to be attributed to the union of several causes, once potent for both evil and good, but now powerless for either except to that limited extent in which the accident of being first in the field in manufactures, will for a time secure to Manchester compensating advantages in competing with other places better suited for the effective prosecution of the cotton trade. Already changes are being made in machinery to suit short staple cottons, which is a well-advised step in the right direction; but it would be wiser still, before laying out money even to that limited extent necessary for such changes, to consider the propriety of resorting to water-power along with these, so that the losses attaching to such alterations may be compensated by the gains arising from the cheaper production, more particularly because the necessity of having to adopt the water-power in self-defence will soon be unavoidable, therefore these changes will not be permanent. All this, however, implies that Ireland's opportunity should be improved even by English manufacturers, and it is so much the better for both countries that it is so. Before money is laid out in any undertaking, it is necessary rather to examine the grounds of safety than to rush after a prospect of profits, only likely of realisation because others have made money in similar businesses before.

We are justified, moreover, by facts of common observation, in saying that no better reason than that of following others without questioning who had made money in the trade, could be assigned by most of the Irishmen who are engaged in cotton manufactures in Lancashire. These persons, disregardful of the lesson taught them by the example of the English people in developing their own resources, rushed away from resources well suited for cotton manufactures, and far superior to those they went to cultivate ; and although many of these have made fortunes, their blunder is not the less palpable. Neither can it be said that their present position is anything the less pitiably ludicrous in that, whilst as members of the class of cotton manufacturers they only suffer as others. The English people cannot feel so much ashamed as the Irish people ought to feel, because they suffer only for what they have done, but are not also reprehensible for what they have not done as the Irish are for leaving places they might have occupied at home, with both honour and gain, for those in which they now bear their part in the consequences of the grossest violations of sound commerce.

In the lessons of the cotton crisis are taught the principles on which Ireland's opportunity hangs. The very same cause operating differently, because of the dissimilarity of circumstances, which led to the cotton crisis, has led also to the neglect of Ireland's resources. In the one, too much was staked on American peace often threatening to break up. In the other, too much was expected of agriculture, which was known to depend on markets for produce, which markets depend on the circulation of money, and that on manufactures, as matters are in these countries.

In both causes the common error of let well alone has led to a common evil of desolation in both countries of the people on whose labour the respective interests of manufactures and agriculture depended. Two years have gone by of the Lancashire suffering. Ten or more have passed over the Irish. The English sufferers have been much relieved by efforts in this and other countries above all praise. Donations have been liberally made, but their suffering is greater than any mere supply of food or raiment can meet; and, therefore, no adequate compensation can ever be made by those who have inflicted the wrong on these patient, noble people, to whose very trials other classes are indebted, for the manifestation of virtues never supposed to exist amongst them.

The Irish sufferers have also been often relieved. They, too, are as patient as could be expected in their lot in life; and to these, as well as to the Lancashire operatives, no adequate atonement can be made by those on account of whose folly, in which they had no part, they have been sunk into such calamities. But both evils are remediable. A return to sound principle will restore Lancashire and raise Ireland. The factory operative in one case, and the agricultural labourer in the other, will participate in the very same reforms in commercial action now rendered imperative in England, and the development of manufacturing resources forced on the people of Ireland.

Here, then, Ireland has an opportunity she sought not for. It is one to which no class can lay claim. It has come at a time when all alike feel the need of relief from a general depression, and if improved may yield benefits to all classes beyond any possible estimation. Educational progress, the cessation of agitations, the undesirableness of America for emigrants, the pressure put upon those who raise crops to get them sold, now that English people require less—all combine to urge upon us the opportunity of making what we have done useful, of completing what we have begun, and of taking advantage of passing events for our own good, now as outlets for our unemployed people in other countries are neither so many nor so attractive. But we are so prone to exaggerate our difficulties, depreciate our advantages, magnify the position gained by the English, and find excuses for our further indolence, that with all the inviting character of the opportunity now before us, there is so much to be done, as calls forth the exertion of all classes as such, and of each individual member in every available way.

If that which is purely imaginary be taken from our difficulties, our special advantages be fairly valued, that which is accidental, and is now even doubtful in the claims of England to commercial prestige, be deducted, and that no apology be tolerated which cannot be defended, we will be in a position to improve Ireland's opportunity.

But to meet the case fully, facts must displace fancies, and activity become the normal condition of men in every rank. Classes, as such, must cease to rely on other classes for pushing forward the progress now forced upon us. Every man, in his own sphere, must do his

L

duty, as if there was no other man in all the country alive to the exigencies of the moment.

The columns of broad sheets, already more devoted to practical matters, must become, to a still greater extent, aids to the industrious, and guides to the thoughtful, on even the more minute matters which concern our industrial development. Young men's associations, now an institution amongst us, ought at once to give less attention to politics, poetry, and juvenile displays of eloquence, and take up more fully the study of chemistry, botany, mineralogy, mechanics, book-keeping, and everything practical to the exclusion of all things merely pleasureable. If we will but become a practical people, and allow the principles of a sound commercial polity to guide our efforts after the utilisation of our material wealth, Ireland's opportunity will be rendered, by improvement, more advantageous to our people than their successes so much applauded in America or England.

Besides, Ireland's opportunity being rightly employed, must open the way for trading people in England escaping from the consequences of efforts (if it ever be attempted) to keep the cotton trade in one place, whether for political or other reasons, now that the principles of free trade demand that it should find its way to whatever place those engaged in it can make it the most profitable. Nor will it be any less manifest, provided this occasion be taken advantage of, that moneyed people in America even now desire an outlet from the consequences of trying to keep together a system, by the abuses of which the elements of its own explosion have already been lodged within its heart.

Neither it is to be overlooked that this opportunity implies chiefly the stirring up of a people whose antecedents prove their competency for any emergency, and whose present condition of apathy is more to be attributed to false notions of its cause—happily departing rapidly— than to any unwillingness to be what they ought to be, or meet those demands which every principle of common sense commends to those most inclined to postpone just and inexorable demands of a sound commercial policy.

To embrace the present opportunity, with all its primary advantages to Ireland's trade, implies even more important considerations for that of Great Britain. Ireland would be elevated to a first place amongst commercial nations; but Great Britain would have advantages in the ports necessary for Irish trade, the like of which she never had as yet, and such as no enterprise could ever secure on her own coasts.

THE growth and sale of TIMBER, the utilisation of IRISH RIVERS for manufacturing purposes, the development of PEAT BOG, the further extension of GLASS BOTTLE making, the increase of JOINT STOCK COMPANIES, the stopping of EMIGRATION, the extension of SHIP-BUILDING, the raising of Irish COAL, the possibility of enforcing SANITARY regulations, the progress of ASSURANCE business, with its necessary adjuncts; BUILDING SOCIETIES and CO-OPERATIVE ASSOCIATIONS, as well as the raising of MARBLE and BUILDING STONES; all come under a few general conditions, and being subject in common to the very same difficulties, become associated together, strange as it may seem to persons disposed to look at them abstractly, in a way at once instructive to those who may be interested in but one of them, and encouraging to every one who wishes to see them all fairly dealt with, for the good of Irish trade and commerce. The anomalous condition of the country affects all material resources, keeps back enterprise in every direction, and renders the most desirable reforms either too expensive or impracticable. The way to change this condition is to aim at manufactures, and to attempt their introduction on a large scale. No greater error can exist than the supposition that because the agriculture of Ireland might be made the most profitable of any in the world, that therefore it is to be followed as either the chief, the first, or the only means for improving the condition of our people. The very persons who are ready to go far in supporting organizations for agricultural improvement, do least of all others to facilitate the transit of Irish peat bog, sands, coal, marble, and stones, from their localities (which are from one to fifty miles from the seaboard), so as these could be allowed to compete with like things brought from other countries. We regret, moreover, to say those very people look upon the keeping of these materials in uselessness as a sort of necessity, to prove that though Ireland has all these, yet it is on agriculture she most of all depends. It is true, she must depend on anything, so long as the fashion is to give but one thing fair play at once; but it would nevertheless be clearly the lesser of two evils to allow all the lands to go out of culture, and to let the pastures rot also, than to keep, as it is now, those rich resources of wealth, which a common maltreatment has brought so strangely together in the neglected condition they are at present. It is actually a fact, that TREES wanted in several parts of this country, and some shipping ports in England, cannot be carried a few miles to the seaboard in Ireland for rates to allow of them being sold at the places of consumption, and therefore the demand cannot be supplied, and the trees cannot be sold, and all parties suffer.

But worse than this, marble can be brought cheaper from Italy or Belgium to Dublin than it can be got from Galway, Armagh, Cork, or Donegal. Building stones, peat bog, coal, and other heavy materials, also suffer in like manner, and yet, in face of all these facts, persons who know that there is water-power in our rivers and money in our banks, which, if used by joint-stock companies (if private individuals will not use them), would lead to the introduction of more railways as a necessity of a manufacturing country, still object to anything being done, because their crotchet is that Ireland is an agricultural country, and therefore they do ;all they can to keep it such, and such alone.

The fact is that the normal condition of Ireland suggests manufacturing most, and agriculture least, although that agriculture be so great as to make it a matter too momentous to be overlooked in any progressive movement.

But all these materials, united in misfortune, are in the hands of the same class of persons, and, consequently, suffer in common from the inexcusable and inexplicable obstinacy of those who have them to let or sell. These people do not make known by any public announcement, or any other effective means, that necessary information as to royalties, rents, or purchase-money required, and yet they wonder why persons who might be expected to work them do not find them out. A little reflection would soon enable all such to discover that the reason resolves itself into the fact that they (whose interest it was to advertise) have actually done their best to hide them.

How opposite to this is the praiseworthy spirit manifested by the Italian commissioner to the Exhibition of 1862, who, in his report, page 28, says, in regard to the facilities for cotton culture :—" *We believe that when in a country there exist all the elements for creating a great branch of commerce, the surest method of inviting to the same both native and foreign capital, is publicity.*" But as yet the business of the author who writes on Ireland's resources, is not the mere statement of what these consist of, where they are to be had, or how they may be used, but rather the disagreeable task of exposing the hindrances cast in the way of their development by persons who have it in their power to do so. If, for example, a person skilled in the dressing of marble wishes to carry on that trade in Ireland, he has not only to suffer from the difficulty of carriage, which might, even as manufactures are now in the country, be overcome, but most of all from the impracticability of those who have the quarries, so that unless he is prepared to sacrifice both time and money, he must, in self-defence, deal with some more mercantile persons than Irish proprietors for a supply of such material, and has to send to Italy or wherever else he can get marble, although the very kind he wants is at or near his place of business. In the same counties (Down and Armagh) may be seen Scotch marbles and granites decorating mansions and public buildings, and Irish granites and marble of more elegant hues and more capable of polish, built into pig-stios or broken into paving-stones, or if allowed to get into a dwelling-house wall or that of any other building, whether it is red or gray granite or marble, it is sure to be "slap-dashed" with lime and sand mortar. Such are some of the facts in respect to marble and granite, which exist in this year of 1863, and which has even more strikingly unaccountable manifesta-tions in other counties than those we have named. It is a fact fully attested by specimens in the hall of the Museum of Irish Industry, in Dublin, and fully corroborated by authentications beyond suspicion, that Cork, Kilkenny, and many other counties, possess marble of the finest hues, in a great variety of colours, but for trading purposes, this fact, like that of the water-powers of the Shannon, Corrib, Lee, Barrow, Bandon, Robe, and other rivers, is rendered useless, because there is as yet no medium of intercourse established between those on whose properties these are and those who could use them. There-fore there is yet a most formidable barrier in the way of Irish progress in such manufactures as these materials are required for. But we by no means consider such difficulties a reason why no attempt should be made in the way of development, but rather look upon them as more likely to be useful than otherwise.

Such natural resources as we have grouped together (because of the necessary connection between the development of one and that of all the rest), being scattered over so wide a range, and being also abundant at so many points, renders the present unwillingness of the proprietors to allow them to be worked on equitable terms, the less a hardship, because information on these subjects is now becoming so generally the common lot of all, owing to the spread of education in Ireland, and not as it used to be, the boast of but a few, that some beginnings are likely to be made somewhere, and these whenever or wherever gone about properly, will lead to the owners of all sorts of such things, disposing of them most probably at lower rates than the like were ever offered at before in any country. In the expected reaction there may be a danger of overtrading, although, hitherto, this has not been an Irish failing. Nevertheless, under such circum-stances as the richness of the country's resources being fully exposed to view, it is not unlikely the present apathy, at no far-off period of the future, might be succeeded by a time of unhealthy excitement.

The best preparation for such a danger is to spread intelligent instruction on commercial matters in general, as well as in regard to the prospects of Ireland in particular. Probably no more profitable study could be engaged in by our young men than that of the history of commerce, but particularly in Britain, since the latter part of the past century. In this history, now written extensively in various kinds of agreeable books, may be found those facts, from which any one must, in the exercise of ordinary reasoning, infer principles of universal application, obviously little known in this country, and be urged into action, evidently disregarded by Irish merchants and others concerned in the utilisation of our resources. Whether the value of these be taken as guides for future expectations, or as rules for immediate operations, they cannot, in Ireland's present circumstances, possibly be overrated. This will appear the more fully in face of daily sad occurrences, and which, although they go far to explain the cause of Ireland's backwardness, yet offer no excuse for further delay in making efforts after the further utilisation of our latent wealth. Manufacturers of glass bottles have used Irish sands successfully, and have offered to do so exclusively, as they could be supplied abundantly; but the owners of these, so soon as they found a market, raised prices above those at which the like qualities could be imported, and so were left to discover that commodities are now valueless, which they might have been selling to the augmentation of their rent-rolls. Several cases of this have occurred in different parts of Ireland. So it is also in regard to timber. Persons have gone to expense in travelling, and devoted their time in selecting trees for beetling beams, railway sleepers, pit props, and other purposes, for which there are tens of thousands of a most suitable kind in the west of Ireland; but just when they expected to do business, the cupidity of the owners imposed difficulties, and they had to lose all past expenses, and leave disappointed. Cases of this kind are, happily, rare, but are, nevertheless, too numerous to be overlooked, and have exerted an influence against industrial progress too considerable to be disregarded in accounting for the causes of its backwardness. Persons of skill, energy, and experience, have met with treatment in their efforts to explore with a view to the working of coal beds, and have suffered to no inconsiderable extent from the unwillingness of owners, under whose properties these lie, to allow them to be worked; and yet we find it stated in society, and sometimes pointed in broad sheets, that it is the English government, or English influence, which keeps our coal-pits useless. It cannot be too emphatically stated that it is no such thing, but rather such conduct and its consequences as that to which we have alluded more slightly than we would like to do were more space at our disposal.

With these specimens before us, we do not wonder that people on whose immediate use of their money depends their living do not invest more freely in experiments in these materials; nor do we hesitate to say that it is unfair, against evidence, and not to be admitted in argument, that the extent to which Ireland's resources are worked is any criterion of their value. This will appear more clearly in view of the well-known fact that persons, be they natives or not, who propose in England to do as persons have proposed to do in Ireland, are not only listened to, and treated fairly, but encouraged, supported, and borne with, until they have time to do justice to the case they undertake to make out, or to work the materials they are led to experiment upon.

So long as money is kept out of circulation, and enterprise discouraged, things will remain in this state; nor need it be wondered that persons unused to the luxuries of sanitary reform allow their feelings of sympathy for persons of very limited means to lead them to oppose the adoption of sanitary regulations in towns, evidently in crying need of their beneficial influence. This fact, although it partly accounts for the slowness of reform in sanitary matters, must also be associated with another not only possessing no colour of propriety, but positively an evil which must give way before cleanliness is almost possible in many towns in Ireland. We mean the tenement system prevalent in Dublin, Galway, Cork, Belfast, in fact, in all large towns where it can be resorted to, and also the hovel system common in smaller towns, which are neither more nor less than nests of filth; and were it not that in these any given sums of money laid out would yield as much more than if spent in building houses fit for human residences, the very persons to whom these belong, being often the magnates of the towns they live in, or the municipal representatives of the people they live upon, would cry out against them, and have them swept from the country, as they must be, before sanitary regulations can be enforced and as they would be were building-societies introduced or co-operative associations organised, for counteracting their influence, by offering the poorer classes places of abode in which they could look cleanly, breathe pure air, and observe the common demands of decency. Personal cleanliness depends much on the cleanliness of the house, but on both depend the cleanliness of streets, to cleanse which, as they are now in many parts of Ireland, offer for the sake of manure alone, a speculation for joint-stock companies, which would pay liberally. But the best way to secure sanitary reform is to promote manufacturing extension as a means of putting it in the power of men to earn money to support what is necessary to health and comfort. To supply a better class of dwelling-houses might, however, be made a business in any town in Ireland even as matters are now. Rents are very high, and land for building is very low; building-materials are cheap and abundant—which facts lead to the conclusion that something might be done profitably, and, certainly, we must admit a necessity presses upon us for improving the homes of our people immediately. Lessening the flow of emigration stands closely related to sanitary reform.

But to these must also be added, in the more full accounting for the present condition of the country, the fact that persons who are prejudiced against progress are obviously so because present circumstances afford them most opportunities for getting usurer's interest on their money, or occasions for personal display, or the producing of hacknied stories about failures in Irish enterprises, to frighten young people out of the country so soon as they become fit to be useful in it. Yet in face of all, if we only make a very small effort, all difficulties will be, at least, so far overcome as to convert them into a useful check on too rapid progress, or a salutary test of sincerity and principle. No greater deliverance is, as yet, to be desired, as no higher position would be for our good. In those institutions, which it is to be regretted are so few, which are now turning out so many well-educated young men, all of whom are becoming more practical as a necessity of the times—we have much promise, despite of all ideal fears of failures. These young men, by a superior education, are fitted to raise up new, resuscitate old, and make improvements in existing manufactures to an extent calculated to bring within the range of scrutiny those sentiments by which the difficulties peculiar to Ireland's progress in manufactures are kept up. It is true that much has to be done to place this country on a footing with England in the manufactures requiring those materials which we have united, because of the common slight cast upon them all. Yet it is equally manifest that it is by persons of intelligence of a higher order, and of education of more general kind, Ireland expects to rise in manufactures, than were possessed by those to whom England was indebted for her elevation to the pinnacle of manufacturing greatness. This being the case, it is to be expected that safer foundations will be laid, wiser plans projected, and that if not more rapid, at least more abiding progress will be achieved in Ireland than ever was attained to in England.

We are surely justified in these high expectations in view of the fact that, in every corner of the land, children (even in the lower ranks) of tender years, are being refined by the infant school system, and that, so soon as able, may now, in any kind of school, be brought into contact with the best kinds of industrial instruction in every branch, and also be taught grammar, geography, arithmetic, and book-keeping; and, as an accident of the rival systems of the country, not only may, but must also be taught sound morality; so that persons intended for unskilled employments may be as well educated now as those used to be for mechanical trades or skilled labour of any kind. These facts become a foundation able to bear very aspiring hopes, which is also much consolidated by opportunities afforded for university training, and the practical character of collegiate studies now as compared with former days. Besides, there exists in the Museum of Irish Industry, in Dublin, the centre of a system about which ought to revolve similar means of acquiring knowledge in every part, as is now in Belfast, and some other places, on chemistry, and the more useful sciences by which the development of the country, to be effective, must be governed. This Dublin Museum, possessing, as it does, special advantages, naturally takes the place of finishing the education of the future manufacturers of Ireland, but it ought also to be regarded by auxiliary and preparatory schools conducted on like principles, though not similarly endowed, as an auxiliary, and not a rival of private enterprise in a like direction.

If none others had been produced but those two young men (Pontefex and Wonfor) to whom his Excellency lately alluded, the claims of the Institution would be worthy of the sympathy of all who wished well to Ireland; but the fact is, in India, America, England, and several places in Ireland, those who graduated in this place occupy the highest positions in chemistry, and to the lectures on geography, botany, and other sciences delivered in this place, many persons, now reaping the advantages of scientific knowledge, owe their first impressions, and are indebted for first instructions, which, had they not been offered gratuitously, would probably not have been availed of at all. These advantages, in their effects on the resources of the country give every encouragement to hope their development may soon take place, and that many joint-stock companies may soon be energetically carried out.

In support of the same prospects, both the direction of the water-shed and the peculiar character of the river system of Ireland naturally lead. In the document already quoted ("Report of the Italian Commissioner"), at page 9, it is said, "but the English, before employing coals, turned to account all the water at their disposal." Now this is scarcely correct, and being stated in a form so powerful for good or evil, we take the liberty of saying that no excuse existed for not using the water-powers of Ireland before using the coal so fully—for all commercial reasons, as well as, in fact, she is part and parcel with England in the United Kingdom, and in the sense, moreover, in which England is said to do anything in manufactures, the term is popularly and properly understood to include the three kingdoms. Ireland has her share in the manufacturing operations of the United Kingdom, and deserves to bear a part of the blame for neglect by which thousands sterling per annum were uselessly thrown away in the difference between the cost of water-power and that of steam, which ought to have been saved in the case of every mill built the last twenty years; although up to that time it may have been wiser to pay the cost of steam for manufactures whilst making experiments, which could not be made except where concentration to a certain extent had brought machinists, engineers, and others together into or near one place.

The money of Irish capitalists, the talents of Irish mechanics, and the labour of operatives, all go to make up those manufactures, by a few unfairly looked upon as wholly English. Besides which, Irish lands grew the produce, and to Irish hands its distribution was usually allotted, by which the seats of manufacturing progress in England were supplied with food. Nor has Ireland failed directly or indirectly to share in the benefits, nor can any movement now be made by which this country will be more, and England less, the centre of operations without affording both, as formerly—a fair chance of either obtaining what each may deserve. The origin of the lesser amount of gains falling to Ireland, and a greater to Great Britain, is in the fact that the latter acted more, and the former less worthy of reward.

But it is cheering for the cause of progress, and admonitory as to practical action, to read in page 9 of the report before us:—"*The Swiss manufactures are worked almost exclusively by water ;*" and in another paragraph, "*There exists, moreover, in Italy, an immense quantity of water-power, at present wasted, and we Italians who are so deficient in fuel, ought most seriously to devote greater attention to the matter.*" Irish people, on whom primarily (as we have often said before) the onus of action must depend, ought to bestir themselves, if but to keep pace with the Italians. But not only in manufactures, but as means of communication from place to place, and from many parts to the sea Ireland's rivers may be used extensively and as profitably in assisting the railway system in the transit of heavy goods, as they may be employed instead of the steam-engine in the turning of machinery.

The Shannon, for navigation, and in connection with tramways, might be made as useful in the development of peat bog, coal, iron, slate, sulphur ore, as in working cotton mills. But it must never be lost sight of that each would be more effectually done by both being aimed at. It cannot be said that for either purposes this river is in an unprepared state, and yet after thousands has been laid out upon it, nothing practical is suggested by those who have gone so far, and, therefore, it is in something like the state, even yet, in which a coat would be after a tailor has made the several parts, but did not seem aware that before it could be worn all must be put together. Such, too, is the state of many works well begun and properly carried on, but left unfinished, in Ireland, which accounts for that principle every day illustrated which some have ignorantly called a fatality about Irish affairs. This common error in great things of not finishing properly is also manifest in little things, for it is a fact that in Ireland workmen in many departments of handicraft, even if as well paid, do not feel bound to finish their work as well as they have often finished the like in England, Scotland, or America. This state of things always exists where governments are urged to do what private enterprise alone can do well.

But the Shannon is not the only navigable river, nor is it the only one fit to turn mills. The lakes of the country in all directions offer facilities for both water traffic and manufactures, and as it is unnecessary, where there are so many, to mention any except the most prominent, and as what is wanted now is not so much a knowledge of what Ireland has, as how what she possesses can be utilised, we only mention that if the people possessed of property on the Lee, Corrib, Suir, Barrow, Liffey, Blackwater, and like rivers, were to do by their properties what English owners have done long ago with coal, they would soon find their incomes extensively increased. An objection

has been raised in reference to fisheries in the flow of dye works. It is scarcely worthy of notice, and yet it may be well to say that the cost of iron piping, 6-inch diameter, or earthenware tubes of a like bore, would not exceed £80, and £40 per mile respectively, and would last so long as that by adding 15 to 20 per cent. on their cost to the incidental expenses at the few places where such an extensive measure as the carriage of dye waste to the sea would have to be taken, the whole case would be met. If, however, half such bore would carry off all such refuse, then the cost would only be half that named above.

FARMING AND MANUFACTURES.

THE extent to which the expansion of manufactures in Ireland mostly interests tenant farmers over and above the additional income arising from a larger growth of flax, may be estimated at an average of £1 11s. 3d. per acre of the cultivated area, which, taken in round numbers at 5,500,000 acres, would amount to £8,284,375 per annum.

At present Irish agriculturists suffer a loss on their butter equal to £1 10s., and on live stock £2 per acre of the pasture, or crops used in feeding milch or beef cows, bullocks, sheep, pigs, and horses, as compared with that which they would realize if there was a home market for those, which there would be if there were manufacturers in the country. On oats and wheat also there is a loss of £1 5s. and £1 10s. per acre respectively, for the same reason. In face of all these losses it need not be wondered that farmers in England and in Scotland can pay £3, £4, or even more, per statute acre, for land inferior to those which farmers in Ireland have at £1 10s., £2, £3, and upwards. The profits made in England and Scotland far exceed these made in Ireland in farming, because Dorset, Devonshire, Kendal, and Shropshire butter sell at higher prices than Cork, Waterford, or Belfast, in the English market; and the reason is, that the former has the advantage of being sold nearer the farm, and the other the disadvantage of a cross-channel transit. The respective packages also necessary for the different mode of marketing is decidedly against Irish butter, and obviously in favour of English.

But all these considerations alike resolve themselves into the absence or the presence of home demand, which hangs on manufactures; and, therefore, before the Irish farmer can get as much for his produce, he must be placed in circumstances like those of his British competitor. Irish raised beef, mutton, eggs, fowl, and every other kind of farm produce suffers equally with butter in competition with British raised. On this difference, however, there is no estimate put in making up the abovenamed sum, which represents merely shippers' profits, freights, brokers' commission, and carriage, all which the *Irish farmer, not the English consumer* has to pay.

In view of these facts no one need be astonished that the most scientific farming industriously prosecuted has led to severe losses in several cases of Scotch and English farmers coming to Ireland. Superiority of soil, comparatively low rents, abundant labour (according to work actually done, labour is as dear in Ireland, although the wages *per diem* are cheaper than in England, Scotland, or America), all fail, because of the absence of home demand to benefit the practical farmer of Ireland. It is for those reasons, *plus* the greater natural facilities we have for flax growing, that the farmer ought to encourage manufactures. But we must even go further, and ask AGRICULTURAL ASSOCIATIONS, as such, to concentrate such aids as they feel inclined to give operative agriculturists in offering facilities by every means in their power for the further expansion of all kinds of manufacturing industry.

In addition to those noble expressions of charity, by which the sufferings of the Lancashire operatives has been largely and timeously relieved, there is a natural desire on the part of all classes in England to support the "*hands*" rather than allow them to emigrate, or even be employed in "the mills" by means that would violate the principles *of free trade*. This is to be commended as the proper course for English people to pursue in defence of home interests; but their doing so ought to be an example to Irish people to do likewise in support of their home interests.

Although it would be no excuse for curtailing Irish contributions, much less stopping the expressions of our sympathy for the sufferers in England, yet, if preparations were made in Ireland according to the necessities of our case in providing employment for our own people, we would be able also to offer employment to English operatives in Irish factories at English rates of wages, for though our manufacturers had to pay high for cotton, the saving by the use of water-power, and the advantages of adopting machinery suited to the times, would give them such advantages over English competitors as would enable them to work in the worst times, whatever be the source of cotton supply, or the issues of the American war.

INDEX.

www.ingramcontent.com/pod-product-compliance
Lightning Source LLC
Chambersburg PA
CBHW021705210326
41599CB00013B/1530